Exploratory Data Mining and Data Cleaning

WILEY SERIES IN PROBABILITY AND STATISTICS

Established by WALTER A. SHEWHART and SAMUEL S. WILKS

A complete list of the titles in this series appears at the end of this volume.

Exploratory Data Mining and Data Cleaning

TAMRAPARNI DASU

THEODORE JOHNSON

AT&T Labs, Research Division
Florham Park, NJ

A JOHN WILEY & SONS, INC., PUBLICATION

Published by John Wiley & Sons, Inc., Hoboken, New Jersey.
Published simultaneously in Canada.

For general information on our other products and services please contact our Customer Care Department within the U.S. at 877-762-2974, outside the U.S. at 317-572-3993 or fax 317-572-4002.

Wiley also publishes its books in a variety of electronic formats. Some content that appears in print, however, may not be available in electronic format.

Library of Congress Cataloging-in-Publication Data:
Dasu, Tamraparni.
Exploratory data mining and data cleaning / Tamraparni Dasu, Theorodre Johnson.
p. cm.
Includes bibliographical references and index.
ISBN 0-471-26851-8 (cloth)
1. Data mining. 2. Electronic data processing—Data preparation. 3. Electronic data processing—Quality control. I. Johnson, Theodore. II. Title.
QA76.9.D343 D34 2003
006.3—dc21

2002191085

Printed in the United States of America.

10 9 8 7 6 5 4 3 2 1

Contents

Preface

As data analysts at a large information-intensive business, we often have been asked to analyze new (to us) data sets. This experience was the original motivation for our interest in the topics of exploratory data mining and data quality. Most data mining and analysis techniques assume that the data have been joined into a single table and cleaned, and that the analyst already knows what she or he is looking for. Unfortunately, the data set is usually dirty, composed of many tables, and has unknown properties. Before any results can be produced, the data must be cleaned and explored—often a long and difficult task.

Current books on data mining and analysis usually focus on the last stage of the analysis process (getting the results) and spend little time on how data exploration and cleaning is done. Usually, their primary aim is to discuss the efficient implementation of the data mining algorithms and the interpretation of the results. However, the true challenges in the task of data mining are:

- Creating a data set that contains the relevant and accurate information, and
- Determining the appropriate analysis techniques.

In our experience, the tasks of exploratory data mining and data cleaning constitute 80% of the effort that determines 80% of the value of the ultimate data mining results. Data mining books (a good one is [56]) provide a great amount of detail about the analytical process and advanced data mining techniques. However they assume that the data has already been gathered, cleaned, explored, and understood.

As we gained experience with exploratory data mining and data quality issues, we became involved in projects in which data quality improvement was the goal of the project (i.e., for operational databases) rather than a prerequisite. Several books recently have been published on the topic of ensuring data quality (e.g., the books by Loshin [84], by Redman [107]), and by English [41]). However, these books are written for managers and take a managerial viewpoint. While the problem of ensuring data quality requires a significant managerial support, there is also a need for technical and analytic tools. At the time of this writing, we have not seen any organized exposition of the technical aspects of data quality management. The most closely related book is Pyle [102], which discusses data preparation for data mining. However, this text has little discussion of data quality issues or of exploratory data mining—pre-requisites even to preparing data for data mining.

Our focus in this book is to develop a systematic process of data exploration and data quality management. We have found these seemingly unrelated topics to be inseparable. The exploratory phase of any data analysis project inevitably involves sorting out data quality problems, and any data quality improvement project inevitably involves data exploration. As a further benefit, data exploration sheds light on appropriate analytic strategies.

Data quality is a notoriously messy problem that refuses to be put into a neat container, and therefore is often viewed as technically intractable. We have found that data quality problems can be addressed, but doing so requires that we draw on methods from many disciplines: statistics, exploratory data mining (EDM), databases, management, and metadata. Our focus in this book is to present an integrated approach to EDM and data quality. Because of the very broad nature of the subject, the exposition tends to be a summarization of material discussed in great detail elsewhere (for which we provide references), with an emphasis on how the techniques relate to each other and to EDM and data quality. Some topics (such as data quality metrics and certain aspects of EDM) have no other good source, so we discuss them in greater detail.

EXPLORATORY DATA MINING (EDM)

Data sets of the twenty-first century are different from the ones that motivated analytical techniques of statistics, machine learning and others. Earlier data sets were reasonably small and relatively homogeneous so that the structure in them could be captured with compact models that had large but a manageable number of parameters. Many researchers have focused on scaling the methods to run efficiently and quickly on the much larger data sets collected by automated devices. In addition, methods have been developed specifically for massive data (i.e., data mining techniques). However, there are two fundamental issues that need to be addressed before these methods can be applied.

- A "data set" is often a patchwork of data collected from many sources, which might not have been designed for integration. One example of this problem is when two corporate entities providing different services to a common customer base merge to become a single entity. Another is when different divisions of a "federation enterprise" need to merge their data stores. In such situations, approximate matching heuristics are used to combine the data. The resulting patchwork data set will have many data quality issues that need to be addressed. The data are likely to contain many other data glitches, and these need to be treated as well.

- Data mining methods often do not focus on the "appropriateness of the model for the data," namely, goodness-of-fit. While finding the best model in a given class of models is desirable, it is equally important to determine the class of models that best fits the data.

There is no simple or single method for analyzing a complex, unfamiliar data set. The task typically requires the sequential application of disparate techniques, leveraging the additional information acquired at each stage to converge to a powerful, accurate and fast method. The end-product is often a "piecewise technique" where at each stage we might have had to adapt or extend, to improvise on an existing method. The importance of such an approach has been emphasized by statisticians such as John Tukey [123] and more recently in the machine learning community, for instance, in the Auto-Class project [19].

DATA QUALITY

A major confounding factor in EDM is the presence of data quality issues. These are often unearthed as "interesting patterns" but on closer examination prove to be artifacts. We emphasize this aspect in our case study, since typically data analysts spend a significant portion of their time weeding-out data quality problems. No matter how sophisticated the data mining techniques, bad data will lead to misleading findings.

While most practitioners of data analysis are aware of the pitfalls of data quality issues, it is only recently that there has been an emphasis on the systematic detection and removal of data problems. There have been efforts directed at managing processes that generate the data, at cleaning up databases (e.g. merging/purging of duplicates), and at finding tools and algorithms for the automatic detection of data glitches. Statistical methods for process control (predominantly univariate) that date back to quality control charts developed for detecting batches of poorly produced lots in industrial manufacturing are often adapted to monitor fluctuations in variables that populate databases.

For operations databases, data quality is an end in itself. Most business (and governmental, etc.) processes involve complex interactions between many databases. Data quality problems can have very expensive manifestations (e.g., "losing" a cross-country cable, forgetting to bill customers). In this electronic age, many businesses (and governmental organizations, etc.) would like to "e-enable" their customers—that is, let them examine the relevant parts of the operational databases to manage their own accounts. Depending on the state of the underlying databases, this can be embarrassing or even impossible.

SUMMARY

In this book, we intend to:

- Focus on developing a modeling strategy through an iterative data exploration loop and incorporation of domain knowledge;
- Address methods for dealing with data quality issues that can have a significant impact on findings and decisions, using commercially available tools as well as new algorithmic approaches;
- Emphasize application in real-life scenarios throughout the narrative with examples;
- Highlight new approaches and methodologies, such as the *DataSphere* space partitioning and summary-based analysis techniques, and approaches to developing data quality metrics.

The book is intended for serious data analysts everywhere that need to analyze large amounts of unfamiliar, potentially noisy data, and for managers of operations databases. It can also serve as a text on data quality to supplement an advanced undergraduate or graduate level course in large-scale data analysis and data mining. The book is especially appropriate for a cross-disciplinary course in statistics and computer science.

ACKNOWLEDGMENTS

We wish to thank the following people who have contributed to the material in this book: Deepak Agarwal, Dave Belanger, Bob Bell, Simon Byers, Corinna Cortes, Ken Church, Christos Faloutsos, Mary Fernandez, Joel Gottlieb, Andrew Hume, Nick Koudas, Elefteris Koutsofios, Bala Krishnamurthy, Ken Lyons, David Poole, Daryl Pregibon, Matthew Roughan, Gregg Vesonder, and Jon Wright.

CHAPTER 1

Exploratory Data Mining and Data Cleaning: An Overview

1.1 INTRODUCTION

Every data analysis task starts by gathering, characterizing, and cleaning a new, unfamiliar data set. After this process, the data can be analyzed and the results delivered. In our experience, the first step is far more difficult and time consuming than the second. To start with, data gathering is a challenging task complicated by problems both sociological (such as turf sensitivity) and technological (different software and hardware platforms make transferring and sharing data very difficult). Once the data are in place, acquiring the metadata (data descriptions, business rules) is another challenge. Very often the metadata are poorly documented. When we finally are ready to analyze the data, its quality is suspect. Furthermore, the data set is usually too large and complex for manual inspection.

Sometimes, improved data quality is itself the goal of the analysis, usually to improve processes in a production database (e.g., see the case study in Section 5.5.1). Although the goal seems different than that of making an analysis, the methods and procedures are quite similar—in both cases we need to understand the data, then take steps to improve data quality.

Fortunately, automated techniques can be applied to help understand the data (**Exploratory Data Mining**, or **EDM**), and to help ensure data quality (by **data cleaning** and applying **data quality metrics**). In this book we present these techniques and show how they can be applied to prepare a data set for analysis. This chapter will briefly outline the challenges posed to the analysis of massive data, the strategies for taming the data, and an overview of data exploration and cleaning methods, including developing meaningful data quality definitions and metrics.

Exploratory Data Mining and Data Cleaning, by Tamraparni Dasu and Theodore Johnson
ISBN: 0-471-26851-8

1.2 CAUTIONARY TALES

A first question to ask is, why are data exploration and data preparation needed? Why not just go ahead and analyze the data? The answer is that the results are almost guaranteed to be flawed. More specifically, some of the problems that occur are:

- **Spurious results**: Data sets usually contain artifacts generated by external sources that are of no interest to us but get mixed up with genuine patterns of interest. For example, a study of traffic on a large telecommunications company's data network revealed interesting behavior over time. We were able to detect glitches caused by delays in gathering and transmitting traffic characteristics (e.g., number of packets) and remove such delays from inherent bursty patterns in the traffic. If we had not cleaned the data, we would have included the glitches caused by delays in the "signature usage pattern" of the customer, and would have detected misleading deviations from the glitched signatures in future time series.
- **Misplaced faith in black boxes**: Data mining is sometimes perceived as a black box, where you feed the data in and interesting results and patterns emerge. Such an approach is particularly misleading when no prior knowledge or experience is used to validate the results of the mining exercise. Consider the case of clustering, a method often used to find hidden groupings in the data for tasks such as target marketing. It is very hard to find good clusters without a reasonable estimate of the number of groups, the relative sizes of these groups (e.g., cluster 1 is 10 times larger than cluster 2) and the logic used by the clustering algorithm. For example, if we use a k-means algorithm that initializes cluster centers at random from the data, we need to choose at least 10 starting clusters to detect two clusters that constitute 10% and 90% of the total data set. Starting with fewer clusters would result in the algorithm finding one big cluster containing most of the points, with a few outliers constituting the other clusters.

  Log-linear models (e.g., logistic regression) are another common example of misplaced faith. The models are successful when the appropriate number of parameters and the correct explanatory variables are included. The model will not fit well if too few parameters and irrelevant variables are included in it, even if in reality the logistic regression model is the correct choice. It is important to explore the data to arrive at an appropriate analytical model.
- **Limitations of Popular Models**: Very often, a model is chosen because it is well understood or because the software is available, irrespective of the nature of the data. Analysts rely on the robustness of the models, even when underlying assumptions about the distribution (often the Normal

density) do not hold. However, it is important to recognize that, although classical parametric methods based on distributional and model assumptions are compact, powerful and accurate when used in the right conditions, they have limited applicability. They are not suitable for scenarios where not enough is known about the data or its distribution, to validate the assumptions of the classical methods. A good example is linear regression, which is often used inappropriately, because it is easy to use and interpret. The underlying assumptions of linear effect of variables and the form of error distributions are rarely verified. A random data set might yield a linear regression model with a "reasonable" R-square goodness-of-fit measure, leading to a false confidence in the model.

Even if a model is applicable, it may be difficult to implement because of the scale of the data. Many nonparametric methods, such as clustering, machine learning, neural networks and others, are iterative and require multiple passes over all the data. On very large data sets, they may be too slow.

- **Buyer Beware—No Guarantees**: Many data mining techniques do not provide any goodness-of-fit guarantees. For example, a clustering mechanism might find the "best" clusters as defined by some distance metric, but does not answer the question of how well the clusters replicate the structure in the data. Testing the goodness-of-fit of clustering results with respect to the data can be time consuming, involving simulation techniques. As a result, validation of clustering in the context of appropriateness to the data is often not implemented. The best or optimal model could still be very poor at representing the underlying data. For example, many financial firms (such as Long Term Capital Management) have mined data sets to find similarities or differences in the prices of various securities. In the case of LTCM, the analysts searched for securities whose price tended to move in opposite directions and placed hedges by purchasing both. Unfortunately, these models proved to be inaccurate, and LTCM lost billions of dollars when the price of the securities suddenly moved in the same direction.

  Another frequently encountered pitfall of casual data mining is spurious correlations. It is possible to find random time series that move together over a period of time (e.g., the NASDAQ index and rainfall in Bangladesh) but have no identifiable association, let alone causal relationship. An accompanying hazard is the tendency to tailor hypotheses to the findings of a data mining exercise. A classical example is the beer–diaper co-occurrence revealed by mining supermarket purchase data. However, its not likely that one can increase beer sales by stocking shelves with diapers.

We hope that the cautionary tales show that it is essential that the analyst must clean and understand the data before analyzing it.

1.3 TAMING THE DATA

There are many books that address data analysis and model fitting in which a single approach (logistic regression, neural networks) stands out as the method of choice. In our experience, however, getting to the point where the modeling strategy is clear requires skill, science, and the lion's share of the work. The effectiveness of the later analysis strongly depends on the knowledge learned during the earlier ground work. For an example, the analyst needs to know, what are the variables that are relevant (e.g., for predicting probability of recovery from a disease—vital statistics, past history, genetic propensity)? Of these, how many variables can be measured and how many are a part of the available data? How many are correlated and redundant? Which values are suspicious and possibly inaccurate?

The work of identifying the final analysis strategy is an iterative (but computationally inexpensive) process alternating between **exploratory data mining (EDM)** and data cleaning (improving data quality **(DQ)**). EDM consists of simple and fast summaries and analyses that reveal characteristics of the data, such as typical values (averages, medians), variability (variance, range), prevalence of different values (quantiles) and inter-relationships (correlations). During the course of EDM, certain data points that seem to be unlikely (e.g., an outlier such as an 80-year-old third grader, a sign-up date of 08-31-95 for a service launched in 1997) motivate further investigation. Closer scrutiny often finds data quality issues (a mistyped value, a system default date), which, when fixed, result in cleaner, better quality data. In a later chapter, we discuss a case study related to a provisioning data base where clearing up data problems unearthed by EDM allowed us to significantly simplify the model needed to represent the structure in the data. We note that addressing DQ issues involves consulting with domain experts and incorporating their knowledge into the next round of EDM. Therefore, EDM and DQ have to be performed in conjunction.

1.4 CHALLENGES

Unfortunately, the analyst has to do considerable ground work before the underlying structure in the data comes into focus. Some of the challenges of EDM and DQ are:

- **Heterogeneity and Diversity**: The data are often collected from many sources and stitched together. This is particularly true of data gathered from different organizations of a single "federation enterprise", or of an enterprise resulting from corporate mergers. Often, it is a problem even for data gathered from different departments in the same organization. The data might also be gathered from outside vendors (e.g., demographics). While the combined information is presented to the analyst as a single data set, it usually contains a superposition of several statistical processes. Analyzing such data using a single method or a black box approach can produce misleading, if not totally incorrect results, as will be explained later.

- **Data Quality**: Gathering data from different organizations, companies, and sources makes the information rich in content but poor in quality. It is hard to correlate data across sources since there are often no common keys to match on. For example, we might have information about Ms. X, who buys clothing from one business unit and books from another. If there is no common identifier in the two databases (such as customer ID, phone number, or social security number) it is hard to combine the information from the two business units. Keys like names and addresses are often used for the matching. However, there is no standard for names and addresses (Elizabeth, Liz; Street, St.; Saint, St.; other variants) so that matching databases using such **soft keys** is inexact (and time consuming), resulting in many data quality issues. Information related to the same customer might not be matched, whereas spurious matches might occur between similarly spelled names and addresses.

  Data quality issues abound in data sets generated automatically (telecommunication switches, Internet routers, e-transactions). Software, hardware and processing errors (reverting to defaults, truncating data, incomplete processing) are frequent.

  Other sources of data integrity issues are bad data models and inadequate documentation. The interpretation of an important attribute might depend on ancillary attributes that are not updated properly. For example, "Var A represents the current salary if Var B is populated. If not, it represents the salary upon termination. The termination date is represented by Variable C that is updated every three months." For Var A to be accurate, timely and complete, Var B and Var C should be maintained diligently. Furthermore, interpretation of Var A requires good documentation that is very rarely available. Such metadata reside in many places, often passed on through word-of-mouth or informal notes.

  Finally, there are the challenges of missing attributes, confusing default values (such as zero, i.e. zero revenue differs significantly from revenue whose value is not known that month) and good old-fashioned manual errors (data clerk entering elementary school student profile types age as 80 instead of 08). In the latter instance, if we did not know the data characteristics (typical ages of elementary school children) we would have no reason to suspect that the high value is corrupt, which would have significantly altered the results (e.g., average age of elementary school kids).

- **Scale**: Often the sheer volume of the data (e.g., an average of 60 Gbytes a day of packet flows on the network) is intimidating. Aside from the issues of collection, storage, and retrieval, the analyst has to worry about summarizing the data meaningfully and accurately, trading-off storage constraints versus future analytical needs. Suppose, for example, that to perform a time series analysis we need at least 30 days worth of data. However, we can efficiently store and retrieve only a week's worth at the most. Therefore, computing and storing statistical summaries (averages, deviations, histograms) that will facilitate sophisticated analysis, as well as developing summary-based analyses, are a major part of the analyst's challenge.

- **New Data Paradigms**: The term "data" has taken on a broad meaning—any information that needs to be analyzed is considered "data". Nowadays, data come in all flavors. We have data that are scraped off the web, text documents, streaming data that accumulate very quickly, server logs from web servers and all kinds of audio and image data. It is a challenge to collect, store, integrate and manage such disparate types of data. There are no established methods for doing this as yet.

1.5 METHODS

In this section we give a brief outline of EDM and DQ methods. In subsequent chapters, we will explore these topics in detail.

A typical data set consists of data points, where each data point is defined by a set of variables or attributes. For example, a data point in a hypothetical data set of network traffic might be described by:

$$(source\_IP\_address, destination\_IP\_address,\\ number\_of\_packets\_sent, number\_of\_hops, time\_taken)$$

The above set of variables enclosed in parentheses is called a **vector of attributes**, where each item in the vector represents an aspect of the data point. Each data point differs from the other. Some attributes, such as the IP address, are assigned and are completely known. Variables such as packets_sent and time_taken vary from data point to data point depending on many observable and hidden factors such as network capacity, the speed of the connection, the load on the network and so on. The variability or uncertainty in the values of the attributes can be represented compactly using a probabilistic law or rule represented by f. A well-known example of f is the Gaussian, or Normal, distribution. In a way, f represents a complete description of the data, so that if we know f, we can easily infer any fact we want to derive from the data. We will discuss this aspect more in Section 2.2. Estimating the probabilistic rule f is important and valuable, however it is also difficult. Therefore we break it up into smaller sequential phases, where we leverage the information from each phase to make informed assumptions about some aspect of f. The assumptions are often pre-requisites for more sophisticated approaches to estimating f.

The first phase in the estimation of f is to gather high-level information, such as typical values of the attributes, extent of variation and interrelationships among attributes. For instance, we can:

- Describe a typical value. "A typical network flow consists of 100 packets, lasting 1 second." The actual attributes of most of the flows should be close to these typical values.
- Quantify departures from typical behavior. "Two percent of the flows are abnormally large."
- Isolate subgroups that behave differently. "The distribution of the duration of flows between Destination A and Destination B differs from that of the flows between Destination A and Destination C."
- Generate hypotheses for further testing. "Is the number of packets transmitted correlated with duration?"
- Characterize aggregate movements over time such as "Packet flows between Destination A and Destination B are increasing linearly with time."

1.6 EDM

A good exploratory data mining method should meet the following criteria:

- **Wide applicability**: The method should make few or no assumptions about the statistical process that generates the data. Distributional assumptions (e.g., the exponential family of distributions) and model assumptions (e.g., log-linear) limit the applicability of models. This aspect is particularly important while dealing with an unfamiliar data set where we have no prior knowledge.
- **Quick response time**: When we explore a data set for the first time, we would like to perform a wide range of analyses rapidly, to gather as much knowledge as possible to determine our future modeling course. From an applied perspective where an analyst wants to explore a real data set to answer a real scientific or business question, it is not acceptable for an analytical task to take hours, let alone days and weeks. There is a real danger of the analysis becoming irrelevant and the analyst being bypassed by the decision-makers. Since data mining is typically associated with very large data sets, the EDM method should not be overwhelmed by large and high-dimensional data sets. Note that models which require several passes (log-linear, classification, certain types of clustering) over the data do not meet this requirement.
- **Easy to update**: Analysts frequently receive additional data (data arrives over time, new sources become available, for example, new routers on the network) and need to update or recalibrate their models. Again, many parametric (log-linear) and nonparametric (clustering, classification) models do not meet this criterion.

- **Suitable for downstream use**: Few end-users of the EDM results have access to gigabytes of storage or hefty processing power. Even if computing power is not an issue, an analyst would prefer a small, compact data extract that allows manual browsing and intuitive inferences about associations and patterns. In this context, an interesting by-product of EDM is **data publishing**, where the essence of the raw data is summarized as a compact data set for further inspection by an analyst. (We discuss this in detail in Section 4.3.3.)
- **Easy to interpret**: The EDM method as well as its results should be easy to interpret and use. While this seems obvious, there are methods, like neural networks, that are opaque and hard to understand. Therefore, when given a choice, a simple, easily understood method should be chosen over methods whose logic is not clear.

Sometimes the findings from EDM can be used to make assumptions for choosing parametric methods, which enable powerful inferences based on relatively little data. Then, a small sample of the data can be used to implement the computationally intensive parametric methods.

In this section, we give a brief outline of summaries that we will later discuss in detail. Statistical summaries are used to capture the properties that characterize the underlying density f that generates the data. There are two possible approaches to understanding f. Note that while we make a distinction between these two approaches for expository reasons, they represent different points on the same analytical spectrum and share a common analytical language. Each approach can often be expressed as a more general or particular form of the other. Furthermore, estimates such as the mean, variance and median play an important role in both approaches.

1.6.1 EDM Summaries—Parametric

A **parametric** approach believes that f belongs to a general mathematical family of distributions (like a Normal distribution) and its specifics can be captured by a handful of parameters, much like a person can be identified as belonging to the general species *Homo sapiens* and described in particular using, height, weight, color of eyes and hair. The parameters are *estimated* from the collected data. The parameters that characterize a distribution can be classified broadly as:

- **Measures of centrality**: These parameters identify a core or center of the data set that is typical—parameters included in this category are mean, median, trimmed means, mode and others that we will discuss in detail later. We expect most of the data to be concentrated or located around these typical values. The estimates can be computed easily from the data. Each type of estimator has advantages and disadvantages that need to be weighed while making the choice. For example, averages are easy to compute but are not **robust**. That is, a small corruption or outlier in the data can distort the mean. The median, on the other hand, is robust, in the sense that outliers do not affect it. However, the median is hard to compute in higher dimensions. Note that estimates such as the mean and median are meaningful by themselves in the context of the data, regardless of *f*, and hence play an important role in nonparamteric estimation as well (discussed below).

- **Measures of dispersion**: These parameters quantify the extent of spread of the data around the core. The parametric approach assumes that the data is distributed according to some probability law *f*. In accordance with *f*, the data thins away from the center. The diffusion or dispersion of data points in space around the center is captured through the measures of dispersion. Parameters that characterize the extent of spread include the variance, range, inter-quartile range and absolute deviation from the median, among others.
- **Measures of skewness**: These parameters describe the manner of the spread—is the data spread symmetrically around the center or does it have a long tail in any particular direction? Is it elliptical or spherical in shape?

1.6.2 EDM Summaries—Nonparametric

The second, nonparametric approach simply computes the anchor points of the density *f* based on the data. The anchor points represent the cut-offs that divide the area under the density curve into regions containing equal probability mass. This concept is related to rank-based analysis common in nonparametric statistics. Empirically, computing the anchor points would entail dividing the sorted data set into pieces that contain equal number of points. In the univariate case, the set of anchor points $\{q_i\}_{i=0}^{i=K}$ is the $\alpha \approx \frac{K}{n}$ set of cut-off points of *f* if:

$$\int_{q(i-1)}^{q_i} f(u)du = \alpha, \forall i, \tag{1.1}$$

where $q_0 = -\infty$ and $q_K = \infty$. q_i are called the α **quantiles** of *f* (see Fig. 1.1).

Quantiles are the basis for **histograms**, summaries of *f* that describe the proportion of data that lies in various regions of the data space. In the univariate case, histograms consist of bins (e.g., interval ranges) and the proportion of data contained in them. (E.g., 0–10 has 10% of the data, 10–15 has the next 10%, etc.). Histograms also come in many flavors, such as equi-distance, equi-depth, and so on. We defer a detailed discussion until later chapters.

The nonparametric approach outlined above is based on the concept of ordering or ranking data, that is, α proportion of the data is less than X_α, and

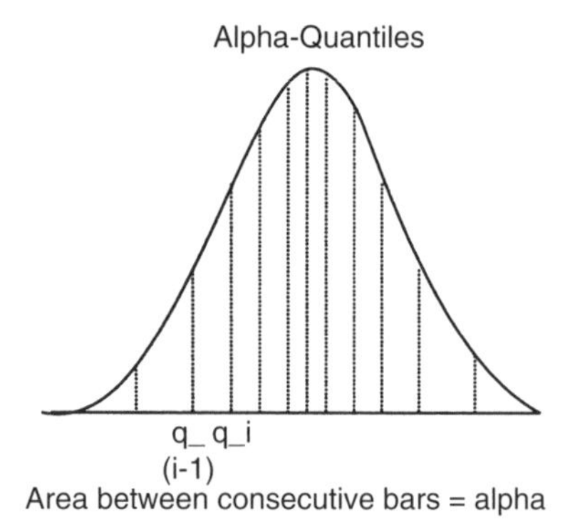

Figure 1.1: α quantiles of f.

so on. In higher dimensions, an analogous concept is **depth**. A data point located deep inside the data cloud has greater depth than one located on the periphery. Examples of data depth include simplicial depth, likelihood depth, Mahalanobis depth and Tukey's half-plane depth and others. Estimating data depth is computationally challenging, involving methods such as convex hull peeling, depth contours, and so on. We will include a detailed discussion in Section 2.9.1.

An important aspect of EDM is to capture correlations and interactions between variables. Many simple measures of capturing bivariate interactions exist, such as covariance, ranked correlation and others. These are easy to estimate but have the same weakness as means, namely, lack of stability. Visual methods include scatter plots, trend charts and *Q-Q* plots. Fractal dimension, mutual information are more complex ways of capturing interaction.

Another important way of capturing interactions is through multivariate histograms. For example, the table below shows that there is a strong association between number of packets and time taken for a packet flow to be transmitted. The numbers in the table represent the proportion of flows that falls in any particular combination of number of packets and duration, such as "Few-Medium" which contains 0.01 of all the flows.

	Short	Medium	Long
Few	0.3	0.01	0
Average	0.08	0.2	0.01
Many	0.01	0.09	0.3

We have created the above bivariate histogram by "discretizing" the numeric variables packets (Few = 0 – 19, Average = 20 – 70, Many = 70+) and duration (Short = Less than 1 sec, Medium = 1 – 5 sec Long = 5+ sec). The combination of the discretized variables results in a **partition** of the data space that has nine **classes** which are exhaustive and non-overlapping.

Partitions of data space are an important way of reducing a large data set into more manageable chunks. Each chunk or class can be represented by summaries of the data points that lie in that class. The summaries are typically orders of magnitude smaller than the raw data. These summaries can be used for further, more sophisticated analysis.

However, it is important to ensure that any given class of a partition consists of data points that are reasonably similar. Otherwise, important differences will be lost in the summarization of the class. For example, if elementary school children and graduate students are included in the same class, then a summary such as "average age" is not representative of either group. Partitions with homogeneous classes have the following advantages:

- As mentioned earlier, the summaries for each class are more reliable and representative.
- Each class is considerably smaller and less complex than the entire data set. Methods suitable for small samples (scatter plots, box plots) could be used on classes that are of particular interest (parts of the network that experience unusual packet loss).
- Representing the data set by a collection of summaries for each class in the partition provides a more detailed (and accurate) understanding of the data set than using one single coarse summary. For example, partitioning the data into two classes, elementary students and graduate students, will give us two average ages for each class (8.9 and 24), rather than a single average age of 17. This kind of a partition, based on an observed attribute (elementary school, graduate school) is called **stratification**, a popular partitioning scheme in Statistics.

The example in the above table is a **rectilinear partition** where the boundaries of the classes are parallel to the axes. A major drawback with creating a rectilinear partition by binning each attribute individually is the exponential increase in the number of classes. If there are d attributes with k bins each, the resulting partition will have k^d classes. Just six attributes with ten bins will result in one million classes! However **data cubes** and **OLAP** software can help the analyst manage this combinatorial explosion (see Section 3.2). Other examples of axis aligned partitions are those induced by classifiers. Clustering methods too induce classes (e.g., each cluster is a class). However such induced partitions are parameterized by the method, so that they do not generalize easily.

Another partitioning scheme, the **DataSphere** or DS, scales well with the number of attributes and is sufficiently general. The number of classes in the partition increases only linearly with the number of variables. The partitioning method consists of dividing the data into **depth layers** around the center (like the layers of an onion) and superimposing **directional pyramids** to capture the axis (attribute) related information. Every layer-pyramid combi-

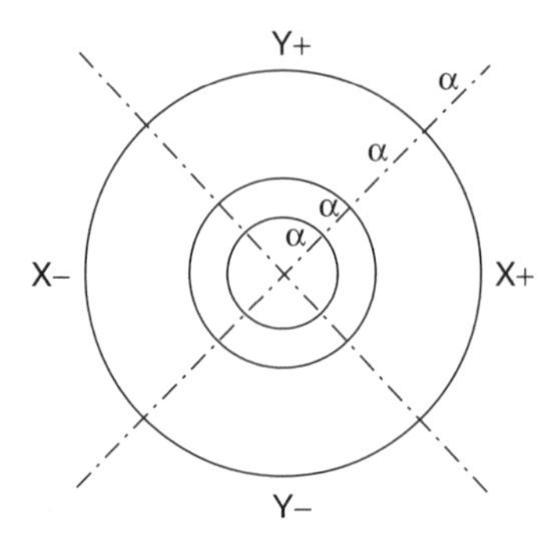

Figure 1.2: A DS partition in 2-D: Depth layers, directional puramids.

nation represents a class in the DS partition (see Figure 1.2). All the points within a class are summarized using aggregates (EDM summaries) that can be combined easily (sums, sum of products, counts). A detailed discussion is in Section 3.4.

Two major uses of partition based summaries computed during EDM are (a) to isolate data glitches and (b) to guide the choice of models for further analyses. Fitting simple nonparametric models within each class of the partition and observing the changes from class to class can lead to an understanding of the nonlinear interactions between attributes. For example, fitting simple survival functions within each class of a partition of the covariates can help us to choose the appropriate proportional hazards model in a survival analysis study. In some cases, such piecewise models can even function as approximations to more sophisticated models.

1.7 END-TO-END DATA QUALITY (DQ)

As noted earlier, data cleaning is an integral part of analysis. In fact,

$$DATA + ANALYSIS = RESULTS,$$

so that the effects of bad data and bad analysis are inseparable. The most sophisticated analyses cannot wring intelligence out of bad data. Even worse, if an analyst is unaware of data glitches, misleading results can be used to make important decisions leading to lost credibility (wrong projections), lost revenues (billing errors), irate customers (billed twice) and sometimes fatalities (incorrect computation of flight paths). Finding data glitches, publicizing them to downstream users and decision makers, and implementing programs to fix the glitches on an ongoing basis should be an integral part of any data quality and data analysis program.

Data quality is a very complex issue, given the innumerable sources as well as the highly domain specific nature of the problems that cause the data glitches. In this section, we briefly outline a comprehensive DQ strategy, with detailed discussions to follow in later chapters. To accommodate the radical changes in the nature of data and what is expected from the data, we update conventional static definitions of data quality to incorporate concepts such as data interpretation, suitability to analysis and availability of metadata to formulate business rules which are dynamic in nature and span multiple systems and processes.

1.7.1 DQ in Data Preparation

Many decisions about data preparation are made during the data processing stage (prior to the first EDM pass). These decisions are made "on the fly" by technicians whose end goal is not necessarily an accurate analysis of the data. The analyst should be involved in these decisions, but frequently is not. As a result, unrecoverable biases are often unknowingly introduced into the data set. For example, consider the choice of default values. While most choices are sensible, sometimes bad defaults are chosen. A negative value (–99999) is a poor default value for an attribute like billed amount, since it is possible to have large amounts credited to a bill.

Another important decision is how to merge different data sources when common keys are either not available or are corrupt. In this situation, domain experts are invaluable. In one of our case studies, the match key was available across three different variables in one data source and across two different variables in the other (for obscure reasons related to the organizational structure). Without the input of domain experts, we would never have identified these keys. In the absence of any common keys, names and addresses are often used. Many tools are available for such name and address matching, for example, Trillium.

Missing values are another source of ambiguity. Discarding all data points that are missing one or more variables can waste a lot of data, and also can introduce unknown biases (all the traffic for destination A to destination B with more than 2 hops is missing). There are many techniques which focus on treating missing values (use typical values, use regression) that we will cover in detail in Section 5.2.2.

1.7.2 EDM and Data Glitches

Partitions are very helpful in detecting glitches. Many data errors are swamped by aggregates. For example, if a small branch of a major company is late in sending the revenues, aggregates such as averages will not be able to detect it.

However, if we break down the data set into a partition, the class in which the branch falls will register a drop, leading to an investigation. We discuss this aspect in detail in our case study on set comparison with DataSphere partitions.

1.7.3 Tools for DQ

No single technique or tool can solve all the data quality issues. Different stages of the process can be tackled using different types of tools. Data gathering and storing can be approached by designing transmission protocols with proper checks and using various ETL tools. Data can be scrubbed and integrated using data browsing techniques, missing value imputation, outlier detection, goodness-of-fit tests and others. In addition, there are tools for dealing with duplicates, name and address correction. Analysis and publishing can rely on EDM and other well-known techniques. The point to note is that a wide range of tools and techniques have to be chosen depending on the data and the task at hand. No single button can be pushed to make the data quality issues disappear.

1.7.4 End-to-End DQ: The Data Quality Continuum

As demonstrated in the previous sections, an effective data mining and analysis program should integrate data quality into the entire lifecycle of the data which we call the **data quality continuum**. Roughly, the stages are:

- **Data Gathering and Data Integration**: Data gathering processes and instruments (software and hardware, others) should be checked frequently to make sure that avoidable errors are weeded out. For example, some systems overwrite certain dates when they run **reconciliation** programs to synchronize databases. The overwritten dates cannot be used for any kind of life cycle or time dependent analyses. In general, it is important to make sure that the data gathered are current, accurate and complete. In addition, the user should be clearly notified and continuously updated of any changes made, and made aware of any non-standard features of the data (e.g., different chunks of the data have different recency) to avoid misleading results and conclusions.

  While trying to integrate data from different sources, a frequently encountered problem is that there is no known **join key** (or match key) to match them on. Clear documentation, binding data to metadata (e.g., XML) and using data browsing to discover join paths are potential solutions.
- **Data Storage and Knowledge Sharing**: Good data models and clear, current documentation are critical for future analysis. Frequently, people in charge of building the data repository are under a time pressure and

fail to create proper documentation. A significant portion of the knowledge, particularly changes in content and convention, is passed on by word of mouth, informally. When the experts leave, the knowledge is lost forever. Therefore, it is important to motivate people to document and share knowledge about the data (the **metadata**), especially business rules which tend to be highly domain specific.

- **Data Analysis**: Our case studies will demonstrate the need for incorporating DQ into analysis. Most analytical techniques start with the assumption that they are given a clean set of data vectors to work with. The analyst should consider potential data glitches, work around them, and caveat the analysis on the possible biases introduced. A frequently encountered problem is adjustment for time lags. For example, while comparing the usage of different customers, we should ensure that the usage records cover the same period. If that is not possible (billing cycle of customer A starts the 15th of every month, whereas the billing cycle for customer B starts on the 25th), the heuristics used for the comparison (e.g., overlapping business days) should be made clear.
- **Data Publishing**: Very few data analysts have the computing power to deal with raw data, therefore, summarized abbreviated versions are published for further analysis, typically on PC platforms, as noted earlier. However, data quality considerations are secondary. Since the data set is summarized, the end user (analyst) is often unaware of propagated errors. Even if he/she notices inconsistent results, the glitches are irretrievably embedded in summaries. Therefore, a strong focus on data quality is particularly important when publishing data for downstream use.

EDM reveals many data glitches. For example, we can use summaries such as averages, variances, and histograms to determine which values are unlikely. Unlikely values or **outliers** are worth investigating since they often represent data glitches. Similarly, time series analysis can be used to detect unusual fluctuations that can be caused by process glitches. For example, sudden drops in revenues could be caused by overlooking the contribution of a biller from some region. Similarly, drops in traffic could be caused by outages or failure in the software that records the traffic.

1.7.5 Measuring Data Quality

Given that data quality means different things in different applications, data quality metrics need to be defined within the context of the problem. Some choices include (a) the increase in usability and reliability of the data, (b) proportion of instances that flow through the process as specified by the business rules, (c) extent of automation, and (d) the usual metrics of completeness, accuracy, uniqueness, consistency and timeliness. A detailed discussion is deferred to Sections 4.2 and 4.5.

1.8 CONCLUSION

We have given an overview of the major aspects of EDM and data cleaning in this chapter. We will elaborate on this theme in the rest of the book, with detailed references and case studies. Our intent is to provide a guide to practitioners and students of large scale data analysis.

CHAPTER 2

Exploratory Data Mining

2.1 INTRODUCTION

Data are collected in many different ways for many different reasons. We are all familiar with sports stats, weather monitoring, census data, marketing databases of consumer behavior, information regarding large galaxies and mountains of data to simulate the motion of subatomic particles. The motivations for analyzing and understanding data sets are equally varied. Census data are used to create high-level summaries and identify large trends. "Those in the age group 35–49 on an average make $100,000 per year," or "The race X, gender Y segment had the highest increase in employment rate." Sports statistics are used to track atypical performances (known as outliers). "Mark McGuire is approaching an all time record for home runs." Customer data are analyzed for finding associations and patterns that can be acted upon. "People who buy candy at the grocery checkout also buy kid's cereals," or "customers who complain about service more than once in a month will most likely switch to a competitor within six weeks of the first complaint." High-level summaries as in the census are easy to compute for almost any data set. But predictions, as in the customer behavior example, require more sophisticated analysis. The choice of the analysis itself is dictated by:

- Our prior experience and knowledge of the data—For example, we might know that only 0.5% of ball players have crossed 60 home runs in a season. So we know that Mark McGuire is exceptional because of our prior experience with baseball performance.
- The quantity of the data—If there were only hundreds of customers, we could use visual techniques to pick out the customer who called two times and then switched to a competitor. However, if there are millions of such customers, establishing the patterns as well as identifying those

Exploratory Data Mining and Data Cleaning, by Tamraparni Dasu and Theodore Johnson
ISBN: 0-471-26851-8

that are inclined toward such patterns becomes very difficult, even with computers.

- Quality of the data—If the people polled in the census lied about their age or income, or they were noted down incorrectly or entered erroneously into the computer, summaries based on such data are meaningless.

In this chapter and the next, we are concerned with exploring large unfamiliar data sets inexpensively, to learn characteristics of the data set. Simple summaries such as typical values of attributes ("a typical person is 68 inches tall, weighs 130 pounds") and the variations in the attributes ("most people are between 60 inches and 76 inches tall, weighing between 100 lbs and 160 lbs") are a good starting point. In addition to characterizing the data, summaries help us to weed out unlikely or inconsistent values that can be further examined for data problems, as discussed below.

Summaries that identify a single characteristic of the data, (such as the average value of an attribute), are called **point estimates**, since they output a single quantity. More complex variations in the data can be captured with summaries such as histograms and Cumulative Distribution Functions (CDFs). Statistical properties of estimates help us to identify summaries that are good for **exploratory data mining (EDM)** (explained below) and data cleaning. Good EDM summaries help us discover systematic structure in data and guide us toward appropriate modeling strategies (e.g., clustering should be used to find groups of customers that buy groceries—veggie lovers, diet fetishists, red meat eaters, couch potatoes, etc.).

In Section 2.2, we introduce an example that will be used throughout the book to informally motivate the concepts of uncertainty, random variables and probability distributions. Our focus is on relating these concepts to exploratory data mining. There are many text books that offer formal and rigorous treatment of these topics. In Section 2.3, we introduce the concept of *Exploratory Data Mining (EDM)* and list the characteristics of a good EDM technique. In Sections 2.4 and 2.5, we discuss summaries (estimates) such as means, variances, medians and quantiles. We outline properties of the summary statistics and identify desirable characteristics of a good summary from an EDM perspective. Such considerations help us to choose rapid and reliable techniques for EDM. Simple estimates like means and medians capture very limited aspects of the variation in the data, so we need more sophisticated summaries for the purposes of EDM. In Section 2.6, we introduce complex estimates like histograms and the empirical cumulative distribution function (ECDF) that capture the variation in attributes across the attribute space. In Section 2.7, we discuss the challenges of EDM in higher dimensions. In Section 2.8, we discuss multivariate histograms that have linear boundaries, parallel to the original variable axes, that is, **axis-aligned**. Since such histograms grow expo-

Figure 2.1: A sleeping Gryphon.

nentially in size as the number of attributes increases, we need more scalable alternatives for fast EDM. Toward this end, we discuss in Section 2.9, *data depth*, its variations and using depth to order data points in higher dimensions. *Depth-based quantiles* can be used for binning in higher dimensions. The next chapter focuses on more sophisticated exploration based on *space partitions* using depth based binning, and capturing complex, nonlinear relationships through EDM summaries. Section 2.9 discusses the role of data depth and multivariate depth which play an important role in multivariate binning. We conclude with a brief summary of the chapter in Section 2.10.

2.2 UNCERTAINTY

Consider the following scenario: A new but stable ecosystem is discovered in the Himalayas. Further, suppose that there are only three species: the mythical beasts Snarks (S), Gryphons (G) and Unicorns (U).

A scientist selects a subset of organisms from the ecosystem to be studied. The subset N of organisms from the ecosystem that is selected and measured is called a **sample** of **size** N from the ecosystem. We would like to infer the properties of the whole ecosystem by studying samples. The scientist painstakingly collects the following data for every member of the sample:

species, age, weight, volume.

The collection of four items above is a description of an individual organism. Each item in the collection is an **attribute** or **variable** whose value gives us information about the organism. We will use the terms variable and attribute interchangeably throughout this book. The values of the attributes vary from organism to organism. For example, two different organisms might be described by the following tuples:

$$(S,4,10,12) \tag{2.1}$$

$$(G,3,9,15). \tag{2.2}$$

We cannot be certain what the value of any particular attribute will be, before we actually catch an organism and measure its attributes. An attribute whose value can vary from case to case is called a **random variable**. The set of all possible values that an attribute (or random variable) can take is called its **support** or **domain**. (Note that the term "support" can have different meanings in other contexts such as association rules in data mining.) In the above example, the support of the attribute species is given by the set S, G, U. The uncertainty in the value of an attribute can be expressed using some function f. For example, if we believe that all three species are equally prevalent, then the uncertainty of which species will turn up as the next data point in the sample is expressed by:

$$\text{chances of species } S = \text{chances of species } G = \text{chances of species } U = 1/3$$

We can represent this in a general form as:

$$f : \chi \to \mathcal{B}, \tag{2.3}$$

where χ is the set of all possible outcomes and $\mathcal{B}$ is the interval [0,1]. For the attribute Species, the set χ is (Snark, Gryphon, Unicorn). The function f is a rule, called a **probability distribution**, that associates a probability of occurrence with every value of an attribute, when the attribute takes discrete values. (We will consider the case of continuous attributes in Section 2.4.2). *A probability distribution represents the uncertainty associated with a particular value of an attribute being observed.*

The probability distribution f is very powerful information, we can answer any question regarding any subset of attributes if we know f. For example, consider the question "What are the chances of catching a Snark whose age is between 40 and 50, and occupies more than 20 units of space?". All we need to do is sum the probabilities (f) of the attributes that lie in the intervals mentioned in the question. So,

$$P((40 \leq A \leq 50) \cap (Sp = S) \cap (V \geq 20)) = \sum_{a=40}^{a=50} \sum_{v=20}^{v=\infty} f(A = a, Sp = S, V = v), \tag{2.4}$$

where the random variable A stands for Age, Sp for Species, and V for Volume and where f ($age = a$, $species = S$, $volume = v$) represents the probability that age is exactly a, species is S and volume is exactly v as specified by the multi-

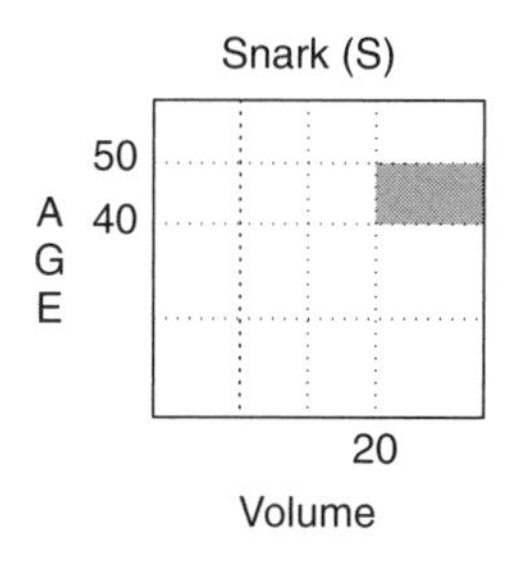

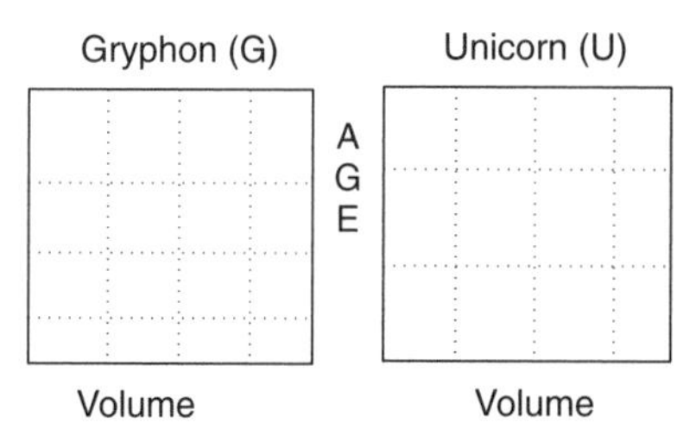

Figure 2.2: A representation of a multivariate *support*. Each box represents a species. The sample organisms that have the attributes Sp = Snark (S), $A \in [40,50]$ and $V \geq 20$ fall in the shaded region. The dotted lines could potentially represent *quantiles* for estimating f.

variate distribution function f, defined in the next paragraph. Figure 2.2 shows a simplified picture.

It is not always possible to express the probability distribution in a simple concise fashion. As $|\chi|$, the size of the set of all possible values for the attribute increases, we cannot list every outcome and the associated probability of occurrence as we did with the species selection. However, there are some popular distributions that can express the probability distribution in the form of a compact mathematical equation. In general, f can be arbitrarily complex, especially if it involves more than one attribute. A **multivariate distribution** represents the probability that a set of attributes takes on a given set of values. (We sometimes refer to a set of attributes as a **vector**.) The probability distribution f may assign the value 0.5 to tuples such as

$$(species = S, age = 30, weight = 10, volume = 15)$$

and the value 0.1 to a tuple like

$$(species = U, age = 3, weight = 5, volume = 55).$$

Intuitively, it is clear that the first set of attributes is five times more likely to appear than the second.

We can think of f as hidden structure in the data, which can be simple and expressed as a compact mathematical expression (e.g., $f(x) = e^{-x}$, the exponential distribution) or arbitrarily complex, not captured by a simple mathematical equation. If f is known, the task of prediction and analysis are very simple. However, in reality, f is seldom known. In order to compute the probabilities like the one in Equation 2.4, we need to guess or **estimate** f, or some approximation thereof, using data. Approximations can range from simple summaries like averages, to classification rules like "if male, aged between 18 and 50, then will see action movie". Building up approximations to the underlying structure f in the data using rapid, scalable techniques is an important task in EDM.

In this book, since the unknown f can be complex, we break the EDM task of estimating f into smaller sequential steps, where we leverage the information from each step to perform increasingly complicated analysis tasks. The first EDM phase in discovering the structure of f is gathering high-level information such as typical values of the attributes, extent of variation and interrelationships among attributes. To illustrate, let us use the ecosystem example. As an initial exploratory phase, we can:

- Describe typical values of attributes. "A typical Snark is 45 units old, weighs 10 units and occupies 16 units of space." The actual attributes of most of Snarks should be close to these typical values.
- Quantify departures from typical behavior. "Two percent of Gryphons have abnormally large weights."
- Identify differences in subgroups. "Snarks and Unicorns have different probability distributions of weight". (See Figure 2.3.) Most Snarks are of Medium weight with relatively few falling in the Light and Heavy categories. In contrast, most Unicorns are either Heavy or Light, with very few weighing-in at Medium. Note that our pictures and explanations are very simplistic (blurring the difference between continuous and discrete attributes) for the purpose of illustration. We will give more rigorous explanations later in the book.
- Generate hypotheses for further testing. "For Snarks, are age and volume correlated?" Or if we had time series information, "Is the size of the population of Unicorns inversely related the size of the population of Gryphons?"
- Characterize aggregate movements in attributes over time. Such information can be used toward building predictive models, such as "Unicorns that have gained weight in the three consecutive time periods are most likely to die."

Typical values also help us to define departures from the normal. For example, if we know that most Snarks weigh 20 Units, if we see that one of the measurements is 200, we should investigate further, to make sure that our measurement was not faulty (device flawed, entered extra 0 by mistake while noting the value). In other words, typical values allow us to identify "abnor-

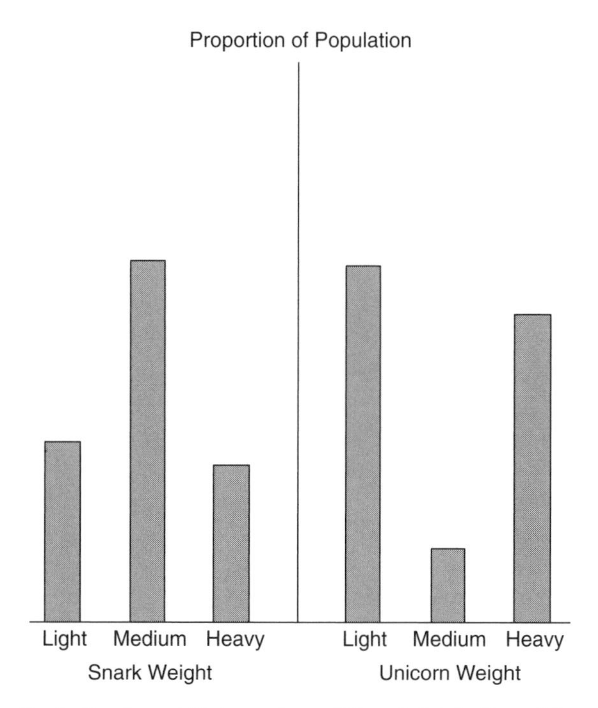

Figure 2.3: Hypothesis: Snarks and Unicorns have different distribution of the attribute Weight.

mal values" which might be **data glitches** as in the above Snark example or which might be genuinely far-out observations ("outliers").

2.2.1 Annotated Bibliography

An overview of probability theory and probability distributions is given in the two classic volumes [44] and [45]. A more introductory and application oriented description is found in [110]. Both the references contain examples of the probability rule f, including discrete distributions such as Binomial and Poisson. Figure 2.1 is from [15].

2.3 EDM: EXPLORATORY DATA MINING

We define **Exploratory Data Mining (EDM)** to be the preliminary process of discovering structure in a data set using statistical summaries, visualization and other means. As mentioned earlier, EDM also reveals unlikely values that are artifacts or inconsistent patterns that frequently turn out to be data problems. Cleaning up data glitches is a critical part of data analysis, which often takes up considerable time, as much as 80% of the total time from the time the data are available to the time to final analysis of the data. EDM helps in *detecting* *the glitches before performing expensive analyses*, avoiding misleading results caused by hidden data problems. Another important aspect of EDM is that it reveals information about the structure in the data that can be used to make assumptions (e.g., f is Gaussian or attribute Y is related to attributes X, U, V in a linear fashion) that facilitate the use of parametric methods (log-linear models, etc.). Such methods enable powerful inferences with strong accuracy guarantees based on relatively little data.

A good exploratory data mining method should meet the following criteria:

- **Widely applicable**: Typically, in a data mining setting, an analyst investigating a new, unfamiliar data set has little or no information about the underlying data. Therefore, a good EDM method should not make any assumptions about the statistical process f (the multivariate distribution) that generates the data. Since we are gathering preliminary information in order to infer some property of f ("How often do Gryphons and Unicorns share the same weight and volume?"), it would be restrictive, if not circular, to make assumptions about f. In fact, the very reason for collecting initial summaries is to help make appropriate distributional assumptions about f (if at all), so that we can use more powerful methods of analysis.
- **Interactive response times**: The purpose of EDM is to quickly investigate several possible methods of analysis and to rapidly eliminate unproductive paths. Therefore, a good EDM method should be fast, even when the number of data points and the number attributes starts increasing. In fact, large data sets are the ones most in need of exploratory techniques for the following reasons. Massive data sets tend to be complex and heterogeneous, so that visual and manual methods are usually not feasible. Although sampling is an option, it is more suited for aggregate inferences about typical instances rather than rare occurrences that are often the target of data mining. Therefore, it is very important that an EDM technique should *scale* well as the data set increases in size, so that an analyst can explore it interactively.
- **Easy to use and interpret**: Methods that require complex transformations of the attributes (such as Principal Components Analysis) are hard to interpret. Most users of EDM might not have the time or expertise to be able to accurately interpret the outcomes. Similarly, neural networks (besides being computationally expensive) are too opaque for a user to feel comfortable with the results. In other words, a good EDM technique should be easy to use and interpret.
- **Easy to update**: Suppose that after we are finished with EDM, we discover that we have missed a group of Snarks, Gryphons, and Unicorns hiding out in a cave. It would be a waste if we had to recompute the summaries all over. Worse, we might have thrown away the raw data since we did not

have enough space to keep them, storing just the summaries instead. But if the summaries and analyses in the EDM are such that we can compute the combined summaries from the summaries of the original set and the data from the creatures hiding in the cave, we can update the values not just this once, but whenever new information becomes available. This is a very critical property for EDM techniques implemented on data that are updated over time, as opposed to a one-time analysis.

- **Easy to store and deploy**: The input and output to EDM techniques should be such that they can easily be stored and deployed. For example, if the summaries produced by the EDM techniques are almost as big as the raw data, then no data reduction or summarization has been achieved.

2.4 EDM SUMMARIES

EDM can be approached in two possibly complementary ways. The first is a "model driven" or **parametric approach**, by assuming a specific functional form for f and estimating the **parameters** that define the function. A parametric EDM approach is useful if we have prior knowledge (nature of the process, previous experience) about the structure of the data. In the ecosystem example, we might assume that the functional form of the distribution of age is **exponential** so that:

$$P(X \leq x) = \frac{1}{\theta} \int_0^x e^{-\frac{u}{\theta}} du$$

and θ is the parameter that we need to estimate in order to know the probability rule f completely. The data would consist of the values of the attribute age, for example, 10, 5, 4, 7, 7, 3, 9, 11, . . . , 6, 9 of N organisms. We would estimate θ by the mean of the N sample values of age.

The second approach to EDM is a "data-driven" or **nonparametric** approach, without any prior assumptions about specific functional form of f or other inter-relationships. Such an approach is used when dealing with new, unfamiliar data sets, where we have no basis for making assumptions.

EDM summaries called **statistics** are computed from the data to capture aspects of the structure in the data. If Z represents the collection of data vectors Z, then we can think of a statistic as a function T that associates a value with every sample Z. Formally,

$$T : S \to \mathcal{R}^d, \tag{2.5}$$

where S is the set of all possible samples Z, $\mathcal{R}$ is the set of real numbers and $\mathcal{R}^d$ is (d)-dimensional space. Examples of statistics T are the sample mean, standard deviation, median and other quantiles. We note that while such point

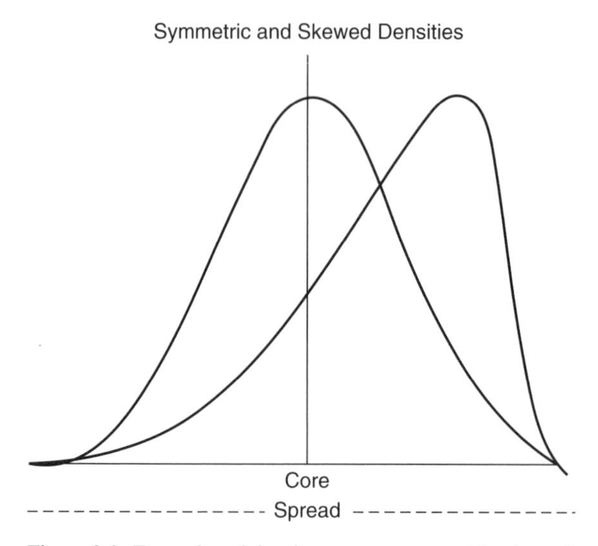

Figure 2.4: Examples of density curves—symmetric, skewed.

estimates when based on the entire data set are too coarse to be valuable, they are powerful EDM summaries when applied to smaller chunks of data and considered together. We will discuss them in detail in the sections ahead. Statistics are an important part of EDM, helping us to construct a navigational map for the structure in the data.

In the example above, the statistic $T(X)$ is the mean age of the organisms in the ecosystem. $T(X)$ is also called an **estimator** of θ. The actual value of $T(X)$ for a given sample X, is called an **estimate** of θ, denoted by $\hat{\theta}$. The hat notation distinguishes an estimate that is specific to a sample (changes from sample to sample), from the true mean θ of the organisms in the ecosystem. The estimate $\hat{\theta}$ gets closer and closer to the real θ as we sample and measure more and more organisms from the ecosystem.

Typically, statistics $T(X)$ constitute EDM summaries which can capture important characteristics of f such as:

- Identify a typical core or center of the attribute distribution that is representative of the population;
- Quantify the extent of spread of the attribute values around the core; and
- Describe the manner of the spread (description of shape, symmetric, skewed). See Figure 2.4.

2.4.1 Typical Values

Certain statistics are designed to measure typical or "central" values of an attribute which are representative of the population in some way. These are called **measures of location** in traditional statistics literature and are an important part of EDM and DQ (Data Quality) analysis. By choosing a handful of representative summaries and using them (instead of the raw data) for further analysis, we speed up the task of EDM considerably. Computing typical values gives us an idea of what to expect and helps us identify "atypical" behavior that can be either due to data glitches or due to genuine outliers. The results are useful either toward data cleaning or toward mining interesting patterns (e.g., high-volume users) that are profitable and not obvious at first glance.

Several statistics have been devised to capture this notion of central or "typical behavior". Each statistic has its own advantages and disadvantages and conceptual motivation. Using several summaries is advisable since each brings out a particular aspect of the data. Often, when used in conjunction, they reveal more about the structure than when used individually. For example, the mean and the median together can reveal information about the skewness in the data, as we will see in Section 2.4.3.

Mean

The **mean** has been traditionally used to represent "typical" values. The mean or expected value of an attribute is the weighted average of all possible values where the weight of any value is its likelihood of occurrence. It can be expressed as

$$E(X) = \mu = \int_{-\infty}^{\infty} u\, f(u)\, du, \tag{2.6}$$

where f is the probability law that governs the distribution of the single attribute X. If the attribute is discrete we merely sum over the product of the value and the corresponding probability. Namely,

$$E(X) = \mu = \sum_{-\infty}^{\infty} u\, f(u), \tag{2.7}$$

where the sum is over all possible values. By convention, probability distributions are represented by $p(x)$ rather than $f(x)$ which is reserved for densities. However, we use a uniform notation since the underlying concepts are similar. Note that not all distributions have means. Heavy tailed distributions, such as the Cauchy density and the Pareto density (for certain parameter values) have infinite means. In addition, categorical attributes such as species can not be "averaged". However, the proportion of a given species is an average analogous to proportion of "heads" in a sequence of coin tosses.

An estimator of the mean μ of an attribute is the **sample mean**. The sample mean is easy to compute and is often the first EDM step. It is given by

$$\overline{X} = \frac{\sum_{i=1}^{N} X_i}{N}, \tag{2.8}$$

where N is the size of the sample. See example in Section 2.4.3 for illustration.

The sample mean varies from sample to sample. Such variations from one sample to the next are called **sampling variations**. Additional samples enrich our understanding of the ecosystem precisely because of this variation. However, we need to be able to bound these variations for the estimates to be of any practical use. For example, it is useful to know that "It is very likely that the mean age of 95 out of 100 samples will lie in the interval [8,12]." Statements such as this are useful for providing **guarantees**. Industrial quality control and process management rely heavily on such guarantees. We can extend this concept to data quality. For instance, if we know that certain subsets of data (female engineers in New Jersey) have certain characteristics (mean income of $150,000) then if we see a subset with a remarkably different value (mean income of $15,000) we can suspect a data glitch and investigate further. In fact, we use a similar approach in our algorithm for automatic detection of glitches in massive data discussed in Section 3.5.

We present below a very terse description of the reasoning behind the guarantees. Please see the bibliography for references to detailed discussions. The guarantees are based on a fundamental result from statistics to construct intervals that enclose the true mean with a high likelihood, called **confidence intervals**. The purpose of such intervals is to associate a measure of accuracy with the estimate, to quantify the reliability of the EDM results. The Central Limit Theorem which (loosely) states that as we choose samples of increasing size N that consist of independently chosen items all with the same distribution f, then the sampling distribution g_N of the sample mean looks more and more like a Normal density, irrespective of what the underlying density f of the sample elements is, as long the mean μ and variance σ^2 of f are finite. (Informally, variance refers to the variation or spread in the values of an attribute. We will define it shortly.) Further, the mean and variance (spread) of g_N are given by μ and $\frac{\sigma^2}{N}$. We can use the sampling distribution of the mean to construct the following enclosing interval for the mean of f.

$$\left[\overline{X} - 1.96\frac{s}{\sqrt{N}},\ \overline{X} + 1.96\frac{s}{\sqrt{N}}\right], \tag{2.9}$$

where the standard deviation s is estimated from the sample by averaging the squared deviations from the sample mean (the standard deviation is defined in detail in Section 2.4.2). The above interval is called a *95% Confidence Interval* for the unknown mean μ of the density f. This implies that we can guarantee that 95% of the means based on samples of size N will lie in this interval. Therefore, samples that yield means outside this interval are "unexpected" and should be investigated for data glitches, abnormal values and other outliers.

The concept of the mean is easily extended to multiple dimensions or attributes, where the multivariate mean is simply the vector of means of individual attributes. For instance, let

$$Z_i = (z_{i1}, z_{i2}, \ldots, z_{id}), \quad i = 1, \ldots, d \tag{2.10}$$

be a sample of multidimensional vectors where d is the number of numeric attributes. Then the mean of the multidimensional vectors is given by

$$\overline{Z} = (\overline{z_1}, \overline{z_2}, \ldots, \overline{z_d}), \tag{2.11}$$

where

$$\overline{z_k} = \frac{\sum_{i=1}^{n} z_{ik}}{n}, \quad k = 1, \ldots, d \tag{2.12}$$

is the mean of individual attributes 1 through d.

The sample mean is attractive since it is easy to compute and conceptually easy to extend to higher dimensions. However, the sample mean is very sensitive to extreme values. It can become corrupted to an arbitrary extent by a single extreme valued data point. This is a major drawback from a data quality perspective, given that dirty data are often the rule rather than the exception. Consider the following scores given to a student on a suite of 10 tests, each scored on a scale of 0 to 100:

$$(95, 90, 93, 98, 91, 90, 98, 97, 99, 9). \tag{2.13}$$

The last score is suspicious since it is so different from the others, while usually test scores are highly correlated. In reality, 95 has been misrecorded as 9, pulling the average score from a legitimate 94.6 to a corrupted value of 86, which might make the difference between acceptance and rejection to a good academic program. While such aberrations have less impact on very large sample statistics, in principle the mean can be distorted irrespective of the size of the sample, by corrupting just a single data point by a very large amount.

One way to address the sensitivity is to trim off a certain proportion of extremely low and extremely high observations. For example, we can drop the scores of 9 and 99 in the above vector of test scores and compute a trimmed mean based on the remaining eight values,

$$M' = \frac{752}{8} = 94, \tag{2.14}$$

which is much closer to the real value of 94.6 than the corrupted mean of 86. Means obtained by snipping off extreme values are called **trimmed means**.

They offer better protection against outliers than dirty data. Clearly, extravagant trimming is a waste of data and is not recommended. Discarding between 2% and 10% (half from each end of the sorted values of the attribute) is acceptable. Trimming more than 20% (total) would not be advisable and would result in the loss of valuable information about the attribute in the atypical portions of its support.

Median

The sample median is another important EDM summary that estimates the center of an attribute distribution. In one dimension, the median is the middlemost value when the sample values are arranged in a sorted order. Roughly half the data points are higher in value than the median. For example, if we list the ages of children in a playground

$$(1, 4, 3, 6, 7, 8, 3, 2, 6), \tag{2.15}$$

then the corresponding sorted set of ages is

$$(1, 2, 3, 3, 4, 6, 6, 7, 8). \tag{2.16}$$

Since there are 9 data points, the median age of the children on the playground is the fifth observation in the sorted set, namely 4. If there are an even number of data points, by convention the median is the average of the two middle observations. With respect to the density f, the median M is the value in the support of the attribute X at which the cumulative probability of occurrence is 0.5. That is,

$$P(X \leq M) = \int_{-\infty}^{M} f(u)du = 0.5, \tag{2.17}$$

where X is the random variable representing the attribute. For a discrete attribute, we simply use the sum over the probabilities associated with the sorted values of the attribute, stopping as soon as the sum reaches 0.5.

Median or "middle" implies ordering attribute values from largest to smallest or vice versa. Since qualitative attributes like species and color cannot be ranked, we cannot define the median for non-ordinal, categorical attributes.

Note that while the mean of a distribution might not always exist (i.e., the value of the formulas 2.7 or 2.8 might be infinite), the median does. Recent algorithms that can compute approximate medians (with error bounds) have made the computation of medians for large data viable.

An important property of the median is its stability. The median does not move drastically away from the center of the data until at least half the data values are changed (as long as the ordering stays the same), as opposed to the mean that can be distorted away from the main body of the data by a single corrupt observation. Clearly, stability is both an advantage as well as a dis-

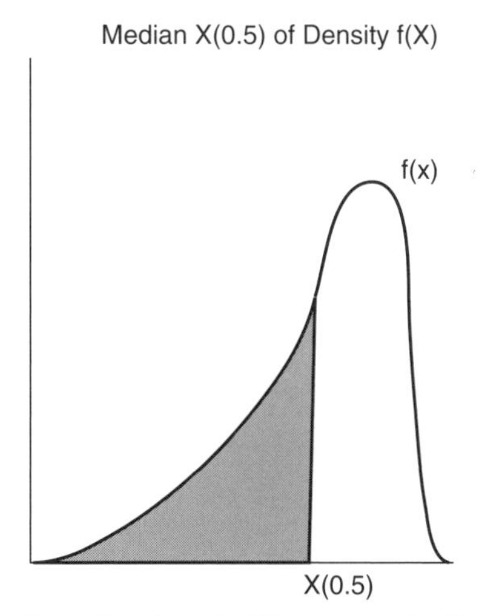

Figure 2.5: Median.

advantage. It is a protection against an occasional aberrant data point. For instance, consider the example of the test scores of the student in Equation 2.13. The median remains between 95 and 97, irrespective of the error in recording 95 as 9, giving the student a fair representation. On the other hand, the insensitivity is a disadvantage since the median might not pick up changes in the distribution until half the data points have been affected. In the example of the children on the playground, let us replace the four older kids by senior citizens to get the new data vector

$$(1, 2, 3, 3, 4, 60, 60, 70, 80). \tag{2.18}$$

The median remains at 4 and does not reflect the change in the age distribution. Such a property is particularly undesirable from a data quality perspective, where we would like to detect glitches before they become widespread or get compounded over time.

The median is hard to generalize to higher dimensions, since sorting by one attribute will disrupt the ordering along other attributes. One option is to compute the median for each attribute and define the multivariate median to be the vector of the attribute-wise medians. Note that the dimension-wise median obtained in this manner is a special case of the trimmed mean where all data points but the middlemost one have been trimmed away. The dimension-wise median would work in special situations where the data are packed in space in some nice way (symmetric, spherical etc.). However, there are instances where it can be quite misleading. Consider the following example with three attributes and three data points:

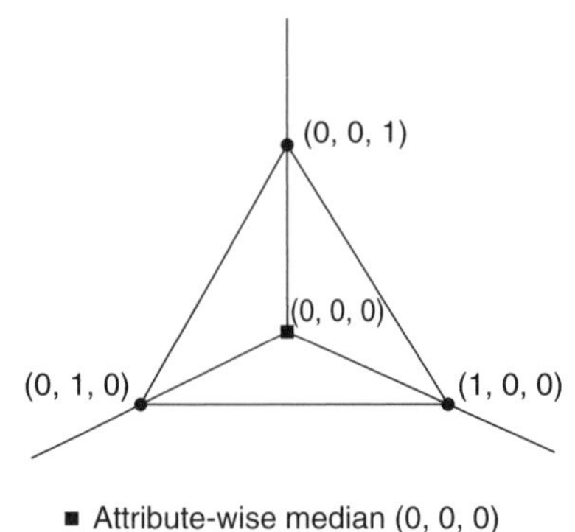

Figure 2.6: Attrbute-wise median lies outside the plane of the data.

$$(1, 0, 0), (0, 1, 0), (0, 0, 1). \tag{2.19}$$

The median of each attribute is zero so that the component-wise median is $(0, 0, 0)$. Figure 2.6 shows that the median computed in this manner is outside a plane that passes through the three points, and is not really a center of this data. On the other hand, the mean (0.33, 0.33, 0.33) lies on the plane and is closer to the concept of a center than $(0, 0, 0)$. Other definitions of the median in higher dimensions require the notion of data depth, which we discuss in detail in Section 2.9.

We will also discuss *quantiles*, generalized versions of the median, and confidence intervals for quantiles, in Section 2.4.2.

Mode

Yet another important EDM summary is the **mode**, the most likely value of an attribute. The mode and its variants (frequency counts) are useful, especially for categorical attributes, where mean and median have no direct meaning, We estimate the mode by choosing the most frequently occurring data point in the sample. Consider the following data vector:

$$(1, 2, 3, 4, 6, 5, 3, 7, 3, 4, 2, 5, 7). \tag{2.20}$$

The data point that occurs most frequently is 3. Finding the mode of the distribution is equivalent to finding the peak of the density f.

However, the mode is not widely used as an indicator of the core or center of a distribution. Its sampling distribution is not well known. Multimodal distributions require combining or choosing among many modes in a meaningful fashion. There has been work recently in finding modes in multivariate and multimodal distributions.

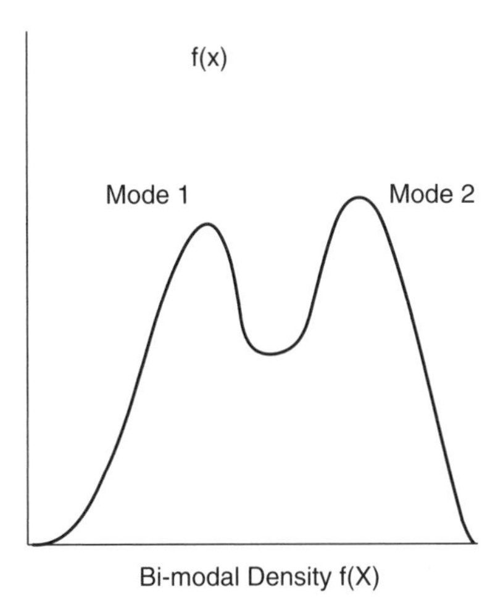

Figure 2.7: Modes of a density.

Summary

We have presented three EDM statistics that estimate typical values of a data set. Each statistic has its advantages and disadvantages, so that some situations call for the use of one type of statistic (median for heavy tailed distributions) while others require a different kind (mode for categorical attributes). Comparison of different measures of location sheds more light on the nature of the distribution of the attribute. For instance, if the mean is much greater than the median, then the distribution is possibly skewed to the right, that is, it has a long, trailing right tail. Hence it is important to compute a variety of statistics even though they represent similar characteristics of the distribution.

2.4.2 Attribute Variation

Some density functions are concentrated over relatively small intervals of the real line (peaked), whereas some are spread over large (flat) ranges. For example, the distribution of age of individuals living in a city is much more spread out (has a greater variation) than the distribution of the age at which individuals first start driving. When a density is spread out, there is a greater likelihood of a large variation in the values of the attribute observed in the data. Summaries that indicate the spread of the data are called **measures of dispersion**.

Measures of spread are important from an EDM-DQ perspective. If we understand how much variation to expect, we can identify abnormal behavior. For instance, if we know that an attribute is most likely to fall between 10 and 40, we will be suspicious if we see a value of 98. This is a powerful means of detecting data glitches as well as rare events that might be of particular interest. In addition, we use measures of dispersion such as the variance, to provide guarantees (or error bounds) of EDM estimates through confidence intervals. If the error bounds or confidence intervals are small, the EDM estimate is more reliable and useful. For example, "the expected arrival time is between 10:00 AM and 10:15 AM" is a more useful statement than "the expected arrival time is between 6:00 AM and 2:00 PM", all other things being equal. Anyone who has had to wait for a cable installation can attest to this.

Other ways of measuring spread are by computing distances between quantiles. Quantiles divide the support of an attribute into intervals that contain equal proportion of data. Therefore, if the deciles (10% of data points lie between consecutive deciles) of the age of Snarks are

$$(5, 6, 10, 12, 14, 16, 18, 20, 25, 26, 30)$$

and that of Unicorns are

$$(2, 6, 10, 14, 18, 22, 24, 26, 30, 40, 50)$$

then we know that the Unicorn age is more spread out than that of the Snarks. The range and the inter-quartile range are constructed from quantiles. Quantiles have other significant applications (histograms, ECDF) that will become apparent later in this chapter.

Another important aspect of dispersion is how attributes vary relative to each other. Understanding inter-relationships between attributes is an important component of EDM that helps simplify and untangle the complex structure in the data (represented by the multivariate density f). By using assumptions such as those of conditional independence (e.g., "if two individuals have no common parent, then their scores on an IQ test are independent of each other"), some nonparametric techniques attempt to construct computationally tractable models. Eliminating linear dependency (collinearity) among attributes is an important part of variable selection (feature selection) for analytical models to eliminate bias and singularity. Attribute relationships are important in data cleaning for finding join paths in the absence of reliable schemas and for validating approximate joins, as we will see in Section 5.3.4.

Attribute inter-relationships can be quantified using many measures such as covariance, contingency tables and Q-Q plots which we will explain in the sections ahead.

Variance: Deviation from the Mean

The **variance** and its square root, the **standard deviation**, play a key role in defining error bounds of expected values. The error bounds, in turn, help detect data glitches, outliers and abnormal values. The variance of an attribute with density f is:

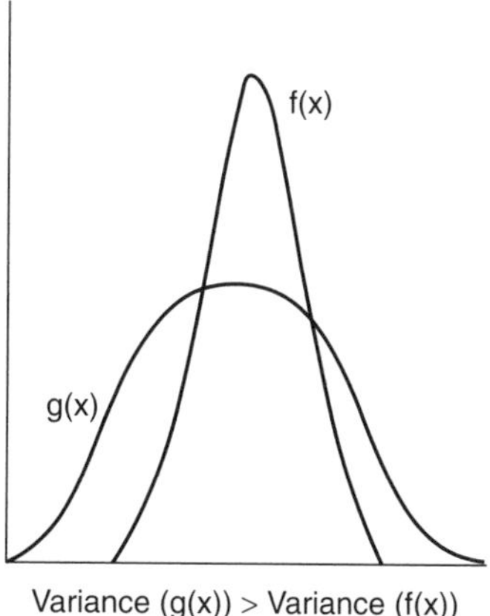

Figure 2.8: Dispersion.

$$\sigma^2 = \int_{-\infty}^{\infty} (u-\mu)^2 f(u)\,du$$
$$= E(X-\mu)^2.$$

The sample equivalent s^2, called the sample variance is the average sum of squares of differences between every data point and the sample mean. That is,

$$s^2 = \frac{\sum_{i=1}^{N}(X_i - \overline{X})^2}{N-1}$$
$$= \frac{\sum_{i=1}^{N} X_i^2}{N-1} - \frac{\left(\sum_{i=1}^{N} X_i\right)^2}{N(N-1)},$$

where N is the sample size. (See Section 2.4.3 for a numerical example.) The mysterious $N - 1$ in the denominator ensures that the estimator $T(X) = s^2$ is unbiased (averages out to the true value), as we explain in Section 2.5. The second step follows upon simple algebraic expansion. From the second step, it is clear that the sample variance can be computed with just one pass over the data by storing the sums and sums of squares of the data points. Note that the variance need not always exist. There are several densities, such as the Cauchy density or Pareto density (for certain parameter values) that have infinite variance. They typically have heavy tails and fall off slowly. Such distributions are seen frequently, for example, in call holding times in the telecommunications industry. Heavy tails imply that extreme values occur more frequently relative to other distributions. For example, we would see a large number of users who tie up resources even when they are not active. From the service provider's perspective, such users are a burden since they consume a disproportionately large chunk of the company's resources compared to the revenue they generate. An easy way to detect heavy tailed distributions is to compare them against a known benchmark using *Q–Q* plots (seen in Figure 2.12) discussed in Section 2.4.4. For instance, in one of our case studies, we noticed that the durations that were conventionally assumed to be exponential were much more heavy tailed than any reasonable exponential distribution. A Pareto assumption was more appropriate. (See Figures 2.11 and 2.12.)

The variance generalizes well to multivariate observations. Let (X,Y) be a pair of attributes, say Z_i and Z_j, of a larger multivariate vector of attributes, $Z = (Z_1, Z_2, \ldots, Z_d)$. By convention, the variance in a multivariate distribution is represented by Σ, called the **dispersion matrix**. The diagonal elements are component-wise variances of the individual attributes. The off diagonal elements are called **covariances** between pairs of attributes. The covariance and the dispersion matrix are measures of association between two attributes which we discuss in the Section 2.4.4 on attribute associations.

Absolute Deviation

The Median Absolute Deviation (MAD) is yet another way of measuring the data spread. For a given sample, the MAD is usually taken from the median. The motivation for MAD lies in the need for robustness. We will see in the chapters ahead (space partitioning, regression) that the measure of dispersion plays an important role in rescaling attributes to establish comparability among them. Otherwise attributes with high variability will dominate the analysis. In this context, it is very important that the estimator of dispersion be robust and representative, and not be sensitive to extreme values. The MAD is given by

$$MAD = median|X - \xi_f(0.5)|, \tag{2.21}$$

where $\xi_f(0.5)$ is the median of the probability distribution f of the attribute X. The sample estimate of the MAD is obtained by using the sample equivalents of the median of f and the median of deviations.

Consider the example of Weight and Age in the example in Section 2.4.3. Observe that:

$$M_a = 11.5$$

$$M_w = 10.5$$

$$MAD_a = median\,(0.5, 1.5, 1.5, 0.5, 3.5, 2.5) = 1.5$$

$$MAD_w = median\,(0.5, 4.5, 0.5, 2.5, 3.5, 0.5) = 1.5.$$

There are other geometric considerations that make the MAD attractive.

Quantiles

Consider a point $\xi_f(q)$ in the support of a numeric attribute X (f refers to the probability distribution of X) such that

$$P(X \leq \xi_f(q)) = q,$$

where f can be dropped from the notation if there is no ambiguity about the probability distribution in question. $\xi_f()$ is called the **quantile function** of f. Given any probability q, ξ_f associates a value X of the attribute, such that the cumulative probability using f exceeds q for the first time as we traverse from the lower end of the support of the attribute X to the higher end. As an example consider a distribution f where each of the 10 values

$$(0, 1, 2, 3, 4, 5, 6, 7, 8, 9)$$

are equally likely. Then,

$$\begin{aligned}
\xi_f(q) &= 0, \quad q \in [0, 0.1] \\
\xi_f(q) &= 1, \quad q \in (0.1, 0.2] \\
&\dots \\
\xi_f(q) &= 9, \quad q \in (0.9, 1.0].
\end{aligned}$$

More generally, consider an attribute that varies continuously, such as weight measured on an ultra-precise scale. The transition from discrete to continuous can be thought of as allowing the attribute to take all possible values. As the number of values that an attribute can take become closely packed, f looks smoother and ultimately becomes a continuous curve. Intuitively, since the attribute can take all possible values in an interval (or union of intervals) the probability is spread among infinitely many values. So instead of associating probabilities with discrete values of attributes such as 1 or 0, we associate probabilities with intervals of values. The probability is higher in certain regions of the interval and lower in others. This is indicated by a continuous function f, which is now called a **probability density** rather than distribution. The peaked or high parts of f correspond to areas where the values of the attribute are very likely. Flatter parts of f that are close to the horizontal axis correspond to values of the attribute that are unlikely. The probability associated with any interval of attribute values is obtained by finding the area under the curve in that interval which is equivalent to "adding up" the probabilities in that interval. The quantiles for the continuous attribute are the set $\{q_i\}_{i=0}^{i=K}$ such that:

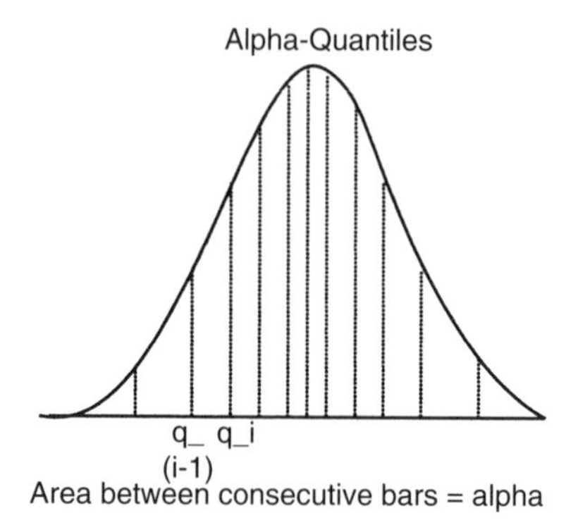

Figure 2.9: Quantiles divide support of an attribute into segments of equal probability.

$$\int_{q_{(i-1)}}^{q_i} f(u)\, du = \alpha, \forall i, \tag{2.22}$$

where $q_0 = -\infty$ and $q_K = \infty$ and $\alpha \approx \frac{1}{K}$. The collection q_i is the set of **α-quantiles** f. The α-quantiles divide the support of X into intervals such that the area under f for each interval is equal, given by α. (See Figure 2.9 for an illustration.)

Empirically, quantiles are estimated by dividing the sorted data set into pieces that contain equal number of points. Conversely, every sample point is a quantile estimate, corresponding to the proportion of data points less than it.

That is, for the sample $X_1, X_2, \ldots, X_i, \ldots, X_n$, we define **indicator variables** $I_i(X)$ such that

$$I_i(X) = 1, \text{ if } X <= X_i$$
$$I_i(X) = 0, \text{ if } X > X_i.$$

As the name implies, indicator variables indicate whether a certain condition is fulfilled or not by taking on the binary values 1 and zero respectively. If

$$\hat{q} = \frac{\sum_{j=1}^{n} I_i(X_j)}{n}$$

is the estimated probability that $X \leq X_i$, then $X_i = \widehat{\xi_f(\hat{q})}$ is the estimate of the q-quantile $\xi_f(q)$ of f. There are many convergence results regarding $\widehat{\xi_f(\hat{q})}$ and $\hat{q}$ that we will not consider here. In essence, we use the sample to compute quantities (probabilities, proportions, quantiles) to mimic the parameters that characterize f.

Analogous to the confidence interval for the sample mean, we need to guarantee the accuracy of the sample median as a stand-in for the median of the density f. By convention, the median of the density is denoted by $\xi_f(0.5)$, where ξ_f is the quantile function. The sampling distribution of the median is well understood when the functional form of f is known. When the underlying data distribution f is not well-known, we can construct a confidence interval for the median using the method below, which we present without proof. Note that we are focusing on the value of the median. (Another approach would be to obtain a confidence interval for the rank of our estimated median as in "we expect the rank of X_m to lie in the interval [0.45, 0.55] with probability 95%" when we repeat the exercise many times.)

- Let $X_1, X_2, \ldots, X_n$ be the sorted values of a sample of n observations for a given attribute. (This is a univariate argument).
- Let X_m be the middlemost observation, the sample median.
- Let $B = [X_{m-1}, X_{m+1}]$ be the interval defined by the neighboring quantiles of the sample median. Then B is a confidence interval for the population median $\xi_f(0.5)$. In general, we can choose any interval $B = [X_{m-k}, X_{m+k}]$, where k should be such that $m + k <= n$ and $m - k >= 0$. While larger values of k result in a greater likelihood or confidence of B containing the median, the confidence intervals get wider and less useful. As an extreme example, $[X_0 = 0, X_n = 1000]$ has a higher confidence of containing the median, but $[X_{m-1} = 456, X_{m+1} = 501]$ is "tighter" and more informative.
- (1) Compute the proportions associated with $X_{(m-1)}$ and $X_{(m+1)}$, namely, $\hat{p}'_L = \frac{m-1}{n}$ and $\hat{p}'_U = \frac{m+1}{n}$, respectively.
- (2) The probability with which $\xi_f(0.5)$ falls in B is equivalent to the probability that the Normal variate $N\left(0.5, \frac{(0.5)(1-0.5)}{n}\right)$ with mean 0.5 and variance $\frac{(0.5)(1-0.5)}{n}$ falls in the interval

$$[\hat{p}'_L, \hat{p}'_U].$$

This probability

$$N\left(0.5, \frac{(0.5)(1-0.5)}{n}\right) \in [\hat{p}'_L, \hat{p}'_U]$$

represents the confidence associated with the interval $[\hat{p}'_L, \hat{p}'_U]$. We can replace 0.5 with q for computing a confidence interval for a general q-quantile $\xi_f(q)$.

There are several statistical packages (e.g., SAS) that will compute confidence intervals for quantiles and ranks.

Range of Values

Another class of measures of dispersion is based on the range of values of an attribute. The **range** of an attribute is the difference between the maximum and minimum values of an attribute observed in the sample. The range is a very intuitive concept, making it an attractive estimator of dispersion. If X_1, $X_2, \ldots, X_N$ is the sample then the range R is:

$$R = max_i(X_i) - min_i(X_i), \quad i = 1, \ldots, N. \tag{2.23}$$

The range is not very robust since it is determined solely by the extreme values which vary wildly from sample to sample. The extreme values are usually a result of rare conditions, such as mistyping. In fact, many analyses require the identification and elimination of extreme values (outliers) to make the results more general and robust. Extreme values are good candidates for data quality checks.

We can use the quartiles to define a more robust version of the range, the **inter quartile range (IQR)**. Let Q_3 be an estimate of the third quartile $\xi(0.75)$ (i.e., 75% of the attribute values are below the third quartile) based on the sample and let Q_1 be an estimate of the first quartile $\xi(0.25)$. (To be precise, we should subscript the quartiles $\xi(.)$ with f and the sample estimates Q_1, Q_2 with the sample size N to show the dependence. However, a simple notation helps in explaining the concepts easily.) The IQR is the difference between the first and third quartiles, given by

$$IQR = \xi(0.75) - \xi(0.25) \tag{2.24}$$

and a guess based on the sample at hand is given by

$$\widehat{IQR} = Q_3 - Q_1. \tag{2.25}$$

Consider the following example:

$$(1, 2, 2, 3, 4, 6, 6, 80). \tag{2.26}$$

In this instance, $N = 8$, $Min = 1$, $Max = 80$, $Q_1 = 2$ and $Q_3 = 6$. Then,

$$R = 80 - 1 = 79$$
$$IQR = 6 - 2 = 4.$$

Table 2.1: Himalayan ecosystem example.

	Subject						Sum	Sum of Squares	Sum of Products
	$i = 1$	$i = 2$	$i = 3$	$i = 4$	$i = 5$	$i = 6$			
Age	12	13	10	11	8	14	68	794	
Weight	10	15	11	14	8	10	68	806	783

The sampling distributions of the range and IQR are not straightforward. The confidence intervals for the quartiles can be adapted for computing the confidence interval for the IQR.

2.4.3 Example

For comparing various estimates and their behavior, consider the simple example of weight and age of Snarks, from the Himalayan ecosystem in Table 2.1. If the subscripts w and a denote weight and age, respectively, then

$$\overline{X}_a = \frac{68}{6} = 11.33,$$

$$\overline{X}_w = \frac{68}{6} = 11.33$$

$$s_a^2 = \frac{794}{6-1} - \frac{(68)^2}{6(6-1)} = 4.67$$

$$s_w^2 = \frac{806}{6-1} - \frac{(68)^2}{6(6-1)} = 7.07$$

$$s_w = 2.66$$

$$R_a = 14 - 8 = 6$$

$$R_w = 15 - 8 = 7$$

$$Med_a = 11.5$$

$$Med_w = 10.5$$

$$MAD_a = 1.5$$

$$MAD_w = 1.5$$

It is interesting that the measures of spread are so varied ranging from a Range of 6 for Weight to a MAD of only 1.5. Each measure of dispersion highlights a different aspect of the distribution. For instance, if the IQR is considerably smaller than the variance (if some additional conditions hold), then we know that the distribution has long tails. Since there is not much additional expense involved, it is useful to collect as many different types of summaries as possible for effective EDM.

2.4.4 Attribute Relationships

During EDM, we are often interested in the inter-relationships between attributes. How are weight and height related to each other? Does weight increase linearly with the amount of calorie intake? If not, how do we summarize a nonlinear relationship between them? How do we determine if two attributes are independent of each other? In this section we discuss simple and fast techniques for exploring relationships between attributes. While the techniques seem radically different in flavor, (e.g., *Q–Q* plots, mutual information) they are based on simple summaries such as counts and sums and use basic principles to infer attribute relationships. It is the easy nature of the summaries and principles, that makes these statistics ideal for use in EDM tasks.

Covariance and Correlation

The **covariance** and **correlation coefficient** are the simplest measures of linear association between attributes. The covariance of two attributes X and Y is given by

$$\begin{aligned} C(X,Y) &= \int_{-\infty}^{\infty} (X-\mu_X)(Y-\mu_Y)df \\ &= E((X-E(X))(Y-E(Y))). \end{aligned}$$

The sample equivalent is given by:

$$\begin{aligned} \widehat{C(X,Y)} &= \frac{\sum_{k=1}^{N}(X_k-\overline{X})(Y_k-\overline{Y})}{N-1} \\ &= \frac{\sum_{k=1}^{N}X_kY_k}{N} - \frac{(\sum_{k=1}^{N}X_k)(\sum_{k=1}^{N}Y_k)}{N(N-1)}, \end{aligned}$$

where $X = Z_i$ and $Y = Z_j$ are two numeric attributes of the data vector $Z = (Z_1, Z_2, \ldots, Z_d)$ and the index k refers to a data point. The *correlation coefficient* of f is a normalized covariance given by

$$\rho = \frac{C(X,Y)}{\sigma_X\sigma_Y}, \tag{2.27}$$

where $C(X,Y)$ is the covariance described earlier, and σ_X and σ_Y are the standard deviations of X and Y respectively.

For the example in Section 2.4.3, the covariance is

$$\widehat{C(w,a)} = \frac{783}{6-1} - \frac{(68)(68)}{6(6-1)} = 2.267.$$

The *dispersion matrix* Σ for the bivariate distribution of Age and Weight is defined to be

$$\Sigma = \begin{pmatrix} V_a & C(a,w) \\ C(w,a) & V_w \end{pmatrix}, \tag{2.28}$$

where V_a is the variance of attribute Age, $C(a,w)$ is the covariance between Age and Weight, and so on. Note that the matrix Σ is symmetric about its diagonal, since

$$C(X,Y) = C(Y,X).$$

The sample estimate of Σ is given by

$$\hat{\Sigma} = \begin{pmatrix} 4.67 & 2.467 \\ 2.467 & 7.07 \end{pmatrix}. \tag{2.29}$$

The sample correlation coefficient is given by

$$\widehat{\rho_{wa}} = \frac{\widehat{C(w,a)}}{s_a s_w} = \frac{2.467}{\sqrt{4.67}\sqrt{7.07}} = 0.439.$$

The correlation coefficient measures the extent of a linear relation between X(i.e., Z_i) and Y(i.e., Z_j) and is normalized to lie between –1 and +1. Large absolute values of ρ imply a strong correlation and small values imply weak correlation. An example of uncorrelated attributes is the number of letters in an individual's name and the individual's age.

If two attributes are independent (e.g., see Figure 2.10 bottom left), then the correlation coefficient should be 0. However, a zero correlation does not imply independence. It merely implies the absence of linear variation in X with Y, and does not exclude a non-linear relationship. For example, consider the tuples (X,Y) given by

$$(-1,1),(1,1),(0,0),(2,4),(-2,4). \tag{2.30}$$

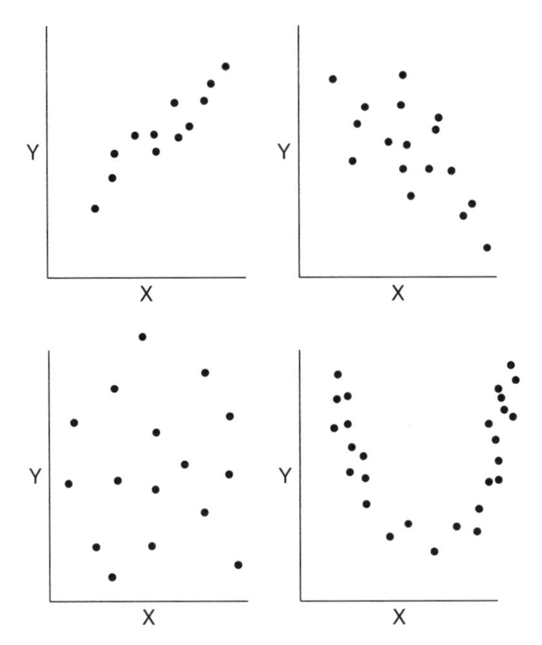

Figure 2.10: Correlation: *Top left*, positive correlation; *Top right*, negative correlation; *Bottom left*, no correlation; *Bottom right*, no correlation but perfect quadratic relationship.

The correlation coefficient is 0, but it is clear that $Y = X^2$, a perfect quadratic relationship (see bottom right, Figure 2.10). While the example we have given is very contrived, the situation where attributes are strongly related but the correlation is 0 occurs frequently in practice, particularly in large data sets. For example, as the number of subscribers of a "closed" service (can be used only if other users have it, e.g., telephone) grows, there is an initial slow increase in usage but once the critical mass of subscribers is reached, there is explosive growth in usage followed by a flatter usage growth period as the service gets saturated. This is a classic nonlinear "**S-Curve**" relationship observed in many data sets.

A positive ρ implies that X and Y move in the same direction. An example of positive correlation is between stock prices and spending. People tend to borrow more and spend more when the stock prices go up. This is called the "wealth effect" (see top left, Figure 2.10). On the other hand, when interest rates go down, the sale of homes tends to go up. There is a negative correlation between interest rates and home sales. A negative correlation is reflected by a negative value of ρ (see top right, Figure 2.10).

Measuring correlations is tricky for categorical attributes like color. However, we can use contingency tables for measuring associations between categorical attributes, as we will see in Section 2.4.4. In the multivariate case,

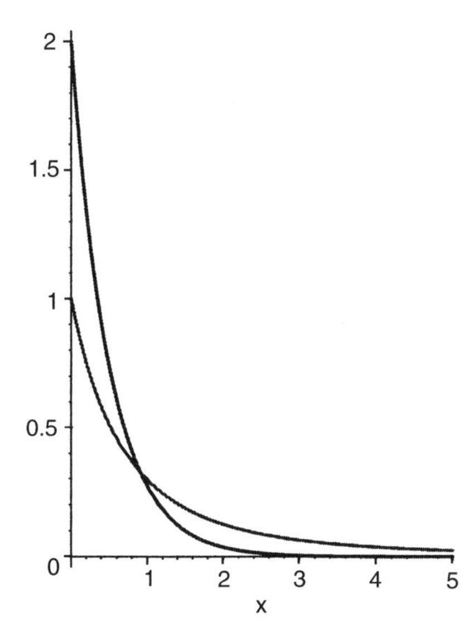

Figure 2.11: Density plot: Exponential versus Pareto.

the determinant of the dispersion matrix Σ is considered a measure of the variation of the multivariate density f.

Q–Q Plots

We have seen that quantiles are good EDM summaries. Quantiles can be used for exploring associations between attributes through *Q–Q* **plots**.

In the most common form of the *Q–Q* plot, the quantiles of a suitably transformed attribute (e.g., [attribute – attribute mean] / attribute standard deviation) are plotted against the quantiles of a well-known distribution such as the Normal with mean 0 and variance 1. The shape of the resulting plot relative to a straight line helps us understand the nature of the attribute distribution.

For example, in Figure 2.11 we have plotted two densities, the exponential and Pareto with appropriate parameters. The Pareto density (the flatter curve) has a longer tail than the Exponential. Now consider Figure 2.12, where we have plotted the quantiles of the exponential versus the Pareto. We can see that the quantiles of the Exponential density are consistently higher than the Pareto (the *Q–Q* plot lies below the straight line $Y = X$) until the very end where the quantiles of the Exponential fall sharply below that of the Pareto, indicating the heavy, long drawn out nature of the Pareto distribution.

Therefore *Q–Q* plots are good EDM tools for understanding the *shape* of an attribute distribution, i.e., whether the shape is skewed to the left, or has a long right tail or has heavy tails.

Q–Q plots can be used to compare marginal distributions (ignoring dependence on other attributes) of attributes. For example, do Snarks and Gryphons have the same age distribution? We can plot the quantiles of the two groups against each other. A straight line would indicate identical distribution and

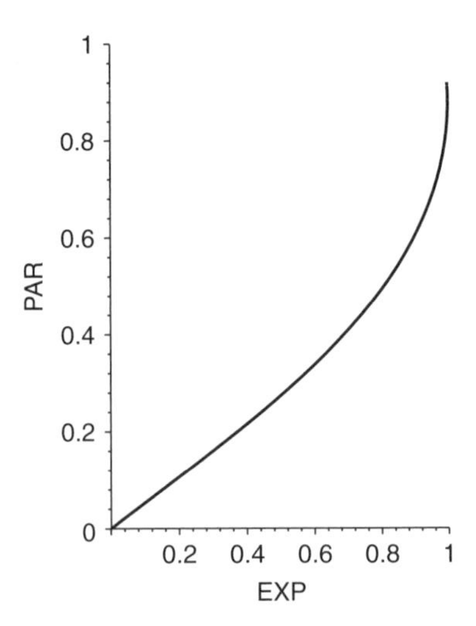

Figure 2.12: *Q–Q* Plot: Exponential versus Pareto. (Axes have been truncated at 1.)

departures can be interpreted in a manner similar to the conventional *Q–Q* plot. This is a very simple but powerful visual EDM technique and convenient since we need to plot just a handful of quantiles. After examining the outcome, more rigorous tests can be performed if needed.

Contingency Tables and Independence of Attributes

A basic concept in probability theory is that of **attribute independence**, which implies that the attributes do not influence or depend on each other. If two attributes are independent, you can multiply their individual probabilities to get the probability of their joint occurrence. We will revisit the notion of independence a little later in the chapter.

A **contingency table** is a cross-tabulation of data with frequency counts for two or more attributes. We can use contingency tables for verifying the independence of attributes. Suppose an attribute X has r categories (e.g., $r = 3$ if X is the attribute species with values Snark, Gryphon and Unicorn in our example) and attribute Y has s categories (e.g., $s = 3$ if Y is the attribute Weight discretized into bins High, Medium, Low) and let n_{ij} be the number of data points in class i of attribute X and class j of attribute Y, then the statistic

$$T(X,Y) = \sum_i \sum_j \frac{\left(n_{ij} - \frac{n_{i.} n_{.j}}{n_{..}} \right)^2}{\frac{n_{i.} n_{.j}}{n_{..}}}, \tag{2.31}$$

called a **chi-square** statistic, can be used to determine whether the attributes X and Y are independent. Here, $n_{i.}$ denotes the sum of all counts in the cate-

Table 2.2: Chi-square test example.

Species	Weight			Row Total
	High	Medium	Low	
Snark	20	30	40	90
Gryphon	20	20	20	60
Unicorn	40	30	10	80
Column Total	80	80	70	**230**

gory i of the X attribute (marginal total of row i) and $n_{.j}$ represents the total of all counts in j category of the Y attribute (marginal total of column j) and $n_{..}$ represents the overall total of all counts. See Table 2.2 for an example and see Section 2.6.1 for related concepts of marginal distributions. $T(X,Y)$ is benchmarked against a well-known distribution called chi-square with $(r-1)(s-1)$ "degrees of freedom" denoted by $\chi^2_{(r-1)(s-1)}$. For the purpose of our discussion, we can think of it as a reference distribution with which we compare the statistic $T(X,Y)$. If $T(X,Y)$ is greater than a particular tabulated value of $\chi^2_{(r-1)(s-1)}$, then we decide that the attributes are NOT independent. If it is less than the tabulated value, we decide that the attributes are independent. The tabulated values are available at various levels of *confidence*. The confidence refers to how reliable the results are. There are several software packages (such as SAS) that will print out the reference chi-square values. We merely mention how to make the decision, without any explanations. A discussion of statistical hypothesis testing is outside the scope of this book.

Consider Table 2.2 to illustrate the chi-square test for independence. The benchmark value of chi-square with $(3-1)*(3-1)=4$ degrees of freedom is **9.48773** corresponding to the cumulative probability 0.95 (also known as 95% confidence level). The chi-square computed according to Equation 2.31 is **24.1295**. Since our computed chi-square exceeds the benchmark, we reject the hypothesis of independence. That is, from our test, we have no reason to believe that the attributes Species and Weight are distributed independently of each other.

The chi-square test is a very versatile test. In Chapter 5, we will discuss using the test to determine the suitability of the data and the analysis to each other.

Mutual Information

Another concept that we can use for testing the extent of associations between attributes is **mutual information**. If two attributes are independent then:

$$P(X=a, Y=b) = P(X=a)P(Y=b),$$

that is, as noted earlier, the probability of joint occurrence is simply the product of the individual marginal probabilities. Mutual information given by

$$I(X,Y) = E_{P(X,Y)}\left(\log \frac{P(X,Y)}{P(X)P(Y)}\right)$$

measures the departure from this expectation, in a slightly different form than the χ^2 test. Note that the expectation is computed over the joint distribution. The closer $I(X,Y)$ is to 0, the less there is in common between the two attributes.

From a computation/estimation perspective,

$$\widehat{I(X,Y)} = \sum_i \sum_j \frac{n_{ij}}{n_{..}}\left(\log \frac{n_{ij}n_{..}}{n_{i.}n_{.j}}\right).$$

In the example in the previous section on the chi-square test for independence, the mutual information works out to be **0.056**. It is hard to interpret this number. However, we can use it to rank attribute pairs in terms of the amount of information they contain about each other for choosing variables to be included in models (feature selection) etc. Note that in actual computation terms, both the χ^2 test for independence as well as mutual information require a two-way table of X-Y counts (n_{ij}), a precursor to bivariate histograms which we discuss later in the chapter.

Fractal Dimension

A significant problem with correlation coefficients is that they only capture the linear relationship between two attributes. However, the relationship between two attributes is often highly non-linear. Mutual information, from the discussion above, measures the expected extent of dependence between two attributes. Computing a **fractal dimension** is an alternative approach which emphasizes the geometric aspect over the likelihood or probabilistic aspect. It is concerned more with the geometric layout of the points in space rather than the density of the points in space.

Suppose that we have normalized the values of two attributes, $X = \{X_1, \ldots, X_n\}$ and $Y = \{Y_1, \ldots, Y_n\}$ so that $0 \leq X_i, Y_j \leq 1$. Let us divide the rectangle containing the data set into r^2 subrectangles, each of which is a $1/r$ square. Let $N(r)$ be the number of rectangles containing at least one (X,Y) point.

What does counting the non-empty rectangles buy us? Suppose that all of the (X,Y) values were concentrated around a single point. Then $N(r)$ would remain almost constant as r increases. Similarly, $N(r)$ will grow linearly with r if the points (X,Y) are concentrated along an (arbitrary) line, and quadratically if X and Y are completely independent. Thus, $N(r)$ gives information about the association between X and Y while making no assumptions about what that association might be. More formally, the **Hausdorf fractal dimension** is defined as

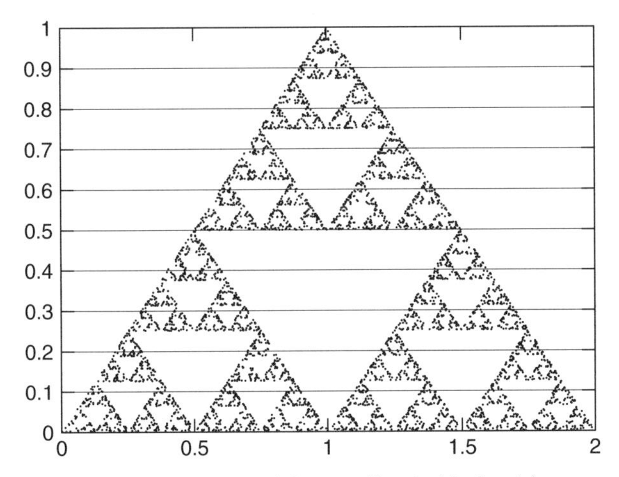

Figure 2.13: A fractal data set (Serpinski triangle).

$$D_0 = \lim_{r \to \infty} \frac{\log(N(r))}{\log(r)}. \tag{2.32}$$

The fractal dimension need not be integral and, in fact, is designed to measure the dimension of "fractal" data sets. In practice, our data sets are of finite size, so that *log*($N(r)$) cannot be larger than n, the sample size, and the limit value of D_0 is 0. However, a plot of *log*($N(r)$) by *log*(r) will have an initial settling-in transient, then a slope of D_0, and then it will flatten out.

Let us consider an example. Figure 2.13 is a two-dimensional Serpinski triangle (a common fractal example), which has fractal dimension 1.585. Figure 2.14 plots *log*($N(r)$) against *log*(r) for this data set, and also the backwards differences (i.e., the slope of the curve). The backwards differences settle into a flat area with a value of about 1.58, closely agreeing with the theoretical fractal dimension. Finally, we note that the fractal dimension can be computed over arbitrary dimension data sets.

2.4.5 Annotated Bibliography

An excellent discussion of statistical estimation, inference and hypothesis testing including summary statistics and estimators, their sampling distributions and asymptotic laws such as the Central Limit Theorem are found in [105]. The book is also a good reference for contingency tables and the chi-square test for independence of attributes. An implementation can be found in SAS software [65].

See [117] for specific discussion of convergence theorems and confidence bounds. See [105] for constructing a confidence interval for the median when

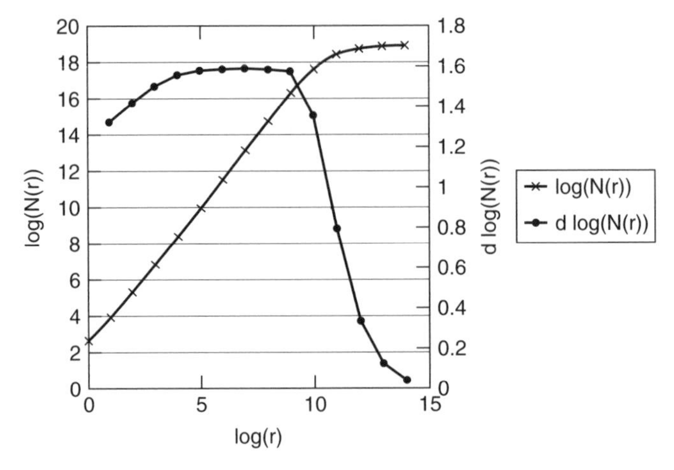

Figure 2.14: Computing the fractal dimension of a 500,000 point Serpinski triangle.

the underlying distribution *f* is known. A measure theoretic approach to probability and convergence is in [8]. See [23] for a discussion of heavy tailed distributions such as the Cauchy and other examples where the mean and higher order statistics could be infinite. The reference also discusses interesting geometric properties of the sample mean. See [60] for a comparative discussion of the mean, the median and trimmed means. The book provides a good overview of exploratory data analysis and robustness properties of estimates like the mean, median, MAD and trimmed mean, and a nice discussion of *Q–Q* plots. A discussion of detecting modes and multimodality can be found in [92]. Rank based tests are discussed in [80]. A theoretical discussion of estimates and their robustness properties can be found in [63]. An interesting account of the breakdown properties of multivariate estimators is in [37]. Mutual information is discussed in [22] and [105], along with related concepts such as entropy.

For a discussion of the rapid computation of approximate medians with error bounds for large data, see [85]. The use of the fractal dimension to determine attribute relationships was proposed in [6]. We used the algorithm described in [4] to generate the Serpinski triangle.

2.5 WHAT MAKES A SUMMARY USEFUL?

EDM often involves collecting summaries and using them for more sophisticated analysis, to speed up computation and reduce storage overhead. Once the summaries are computed, they are the authoritative source of information (e.g., *data publishing*, where manageable, summarized versions of the massive data are published for downstream analysis) and the raw data are typically never accessed again. Therefore, it is important that the summaries be accurate, be general enough to serve as many analyses as possible, be robust and be easy to compute, among other things.

In this book, we try to present different types of summaries that serve complementary needs. Techniques such as fractal dimensions are more geometric than probabilistic. Methods like mutual information, Q–Q plots and piecewise summaries, discussed in the next chapter, are exploratory and sometimes visual in nature. In addition, we also rely on classical summaries that help us provide confidence guarantees, upper and lower bounds and convergence assurance, which are comforting from an analysis perspective. Such assurances are insurance against our results being too unstable or volatile and measure the *accuracy and reliability* of the EDM estimates. We present below some characteristics that are desirable in an EDM summary. Not all of them are applicable to all the summaries that we have discussed. The first set is related to statistical summaries and associated guarantees. The second set of properties are related to computational considerations such as scalability to large multidimensional data sets.

2.5.1 Statistical Properties

We will briefly describe four important statistical properties of sample statistics. Some of these properties were motivated by having too little data (small samples) and might not be an issue in EDM in general. However, when we break up the data into smaller chunks (partitioning), to speed up analysis and make it scale, the properties discussed in this section become important.

Unbiasedness guarantees that the statistic $T(X)$ used for EDM to understand the structure in the data gets closer on an average to the true parameter θ, as we use more data for computing the statistic. For instance, we use a sample mean as an estimate of the true average. We do not use the top 20 data points or the lowest 10%, because if we do the average we compute will have a **bias**. See Figure 2.15, top panel, for an example of a biased estimator as its values drift upwards with increasing sample size. Unbiased estimates are reliably close to the true value of the parameter. However, a trimmed mean has other desirable properties like being relatively immune to outliers and corruption. Therefore, it is important to choose a wide variety of statistics for the purpose of *data publishing* and *data reduction*, where the summaries could be used for further analysis.

The unbiasedness property is expressed formally as

$$\lim_{N\to\infty} E(T_N(X)) \to \theta, \tag{2.33}$$

where $E(T(X))$ is the expected value of $T(X)$, θ is the true parameter that $T(X)$ estimates and N is the number of data points that are used to compute $T(X)$. The limit notation $\lim_{N\to\infty}$ implies that we are choosing progressively larger samples.

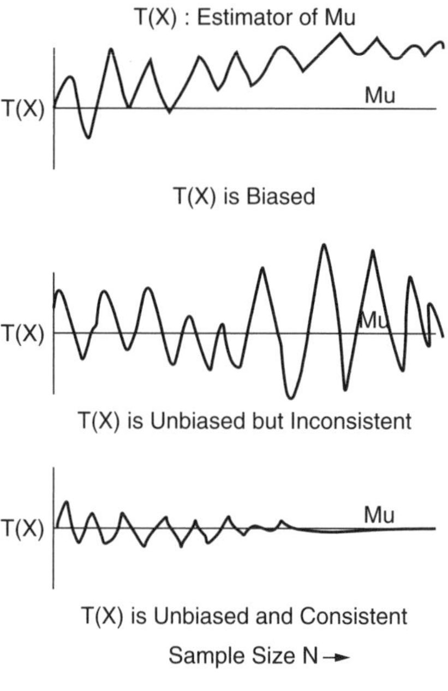

Figure 2.15: Properties of estimators.

An unbiased estimate of the mean of the density f, (regardless of f) is given by choosing $T_N(X)$ to be the sample mean. Recall that $T_N(X)$ is a random variable that varies from sample to sample due to sampling variation.

It can be shown that the sample variance is not an unbiased estimate of the true variance σ^2 of f. The bias $b(s^2)$ is given by

$$b(s^2) = E(s^2) - \sigma^2 = \frac{-\sigma^2}{N}, \tag{2.34}$$

where σ^2 is the true variance of the density f. If σ^2 is large, the sample needs to be quite large to keep the bias down to an insignificant level.

Consistency of $T_N(X)$ implies that the fluctuations in the value of $T_N(X)$ get smaller and become arbitrarily close to zero as more data points are included in the computation of $T(X)$. Again, the motivation is to choose statistics for EDM that are reliable, and do not change wildly when additional data become available. See Figure 2.15, middle panel, for variations that do not get dampened down when the sample size becomes increasingly large. Consistency can be expressed as

$$\lim_{N \to \infty} V(T_N(X)) \to 0. \tag{2.35}$$

It can be shown that the sample mean is consistent, that is,

$$V(\overline{X}_N) = \frac{\sigma^2}{N}. \tag{2.36}$$

The subscript denotes the dependence on the sample size. Clearly, as N becomes large, $V(\overline{X}) = \frac{\sigma^2}{N}$ becomes smaller, ultimately becoming 0.

The first panel in Figure 2.15 shows a biased estimate that drifts away from the true value μ as we incorporate more and more data into the estimate. The second panel shows an estimator that continues to have high variability as the sample size increases. The third panel shows a nice estimator that becomes more and more reliable (low variability) as well as accurate (closer to the true value) as the sample size increases.

Sufficiency is a way to measure the value of a statistic for the purpose of data summarization and reduction. Sufficiency guarantees that, for a given family of densities $\mathcal{F}$, knowing the statistics $T(X)$ is sufficient to have complete knowledge of the density, individual data points provide no additional information. For example, if we are given that f is a Poisson density where

$$P(X = a) = \frac{\exp^{-\lambda a}}{a!}, \quad a = 1, 2, \ldots,$$

with an unknown rate parameter λ, then it is enough to know the sum of the sample observations, (we do not need to know the individual sample values) to replace the parameter in the density and specify it completely. When nothing is known about $\mathcal{F}$, the ordered statistics (i.e., the ordered sample) are the sufficient statistic. The less we know about the family of densities $\mathcal{F}$, finer the granularity of the summaries and lesser the data reduction achieved.

Finally, a statistic $T(X)$ is **efficient** if it has the minimum possible variance. This is a valuable EDM property since it implies that $T(X)$ is the most accurate estimate (least variance) that we can find. In other words, we can bound our guess $T(X)$ by the smallest possible interval. For instance, the statement "The average weight of a Snark is between 10 and 12" is a more informative or stronger statement than "The average weight of a Snark is between 0 and 40". The lower bound of the variance of an estimator $T(X)$ is given by a bound called Fisher's Information Limit. However, the Information Limit need not always be achieved.

Our intention in the preceding discussion is to provide a flavor of the desirable properties that classical estimates have. We can leverage these properties to certify the accuracy of our results. However, such estimates have to be used in conjunction with other estimates that might not have these properties but fulfill other needs of EDM and data cleaning.

2.5.2 Computational Criteria

For EDM, interactive response times are important, hence statistics that take too long to compute are not valuable. Summaries that do not result in considerable data reduction over the original data are not desirable. We might as well keep the original data around instead of wasting resources computing the summaries. Summaries that can be aggregated across different data sets and different pieces of a single data set are important from an EDM perspective. For example, when additional data becomes available, we need to aggregate these summaries for the combined data. Conversely, during EDM, we often need to break the data into smaller pieces (partition), compute summaries and put them back together while building piecewise models to approximate nonlinear relationships (see next chapter). Sums, counts, min, max are all aggregable. A mean without a count is not aggregable.

2.5.3 Annotated Bibliography

A discussion of the statistical properties of sample statistics is in [23] as well as [105]. The latter book has a good discussion of Fisher's information and other efficiency bounds for statistics.

2.6 DATA-DRIVEN APPROACH—NONPARAMETRIC ANALYSIS

In the discussion so far, we discussed an underlying structure f that generates the data, and its characteristics (center, spread) that can be estimated from the data. If we have reason to believe that f has a certain structure (as in Bayesian approaches), the estimates that we have discussed are very useful. When we have absolutely no knowledge or prior experience of the data, we have no starting point to make the model assumptions. We have to use the data as it is, and investigate its characteristics. The characteristics can be explored with simple counts (which values or combination of values occur frequently, histograms), identifying values that co-occur a significant number of times (association rules), deriving rules that can be applied generally (classification, neural networks), identifying co-dependencies among sets of attributes (Bayesian networks), and so on. We can term this approach a *nonparametric* one, since we do not make any prior assumptions about an underlying true distribution f, its parameters or any interrelationships between attributes. While we make the distinction between parametric and nonparametric approaches to underscore the different approaches to data analysis, note that they do share a common framework of concepts and should be considered to be at different ends of an analysis spectrum. For instance, a nonparametric approach can be considered to be parametric with an infinite number of parameters. Additionally, estimates like the mean and the median play an important role in both parametric and nonparamteric approaches.

Table 2.3: Example of a frequency table: age vs. weight.

Age	Weight					Row Total
	0–5	5–10	10–30	30–50	50+	
0–10	3	4	5	2	1	15
10–20	5	10	2	4	0	21
20–30	4	7	2	1	1	15
30+	1	2	0	1	0	4
Column Total	13	23	9	8	2	55

In this section, we will briefly discuss a subset of nonparametric methods that are useful for EDM and DQ. We start with techniques for counting frequencies and simple summaries, such as histograms. We do not discuss well-known techniques like classification, neural networks, Bayesian nets and so on, since there are many textbooks that cover these techniques.

2.6.1 The Joy of Counting

In the previous section we discussed point estimates (estimate a single quantity) like sample means, and their characteristics to summarize coarse properties of the data set as an initial EDM step. However, we need to understand more complex structure in the data such as localized variation of attributes and their distribution in space, intricate nonlinear relationships among attributes and other fine-grained interactions. For instance: How many people in New Jersey travel on a particular stretch of highway between 8:00 AM and 10:00 AM on weekdays? What items are frequently bought together so they can be co-located on grocery shelves?

In the exploratory spirit of EDM (fast, simple, approximate), we can construct a **frequency table** of counts. (We will address the issue of scalability of such a solution shortly.) Consider the two-attribute Table 2.3 based on the Himalayan ecosystem example. The Weight attribute is represented by the intervals 0–5, 5–10, 10–30, 30–50 and 50+ (all values of weight exceeding 50). All organisms whose weight lies between 0 and 5 will be treated as one group with no further distinction in their weights. A similar collapsing is done for the age attribute. The numbers at the intersection of a row and column represent the number of organisms that fall into the corresponding age and weight intervals. There are 10 organisms whose age is between 10 and 20 and weight is between 5–10. The last column represents the row totals and the last row represents the column totals.

The counts corresponding to totals of individual rows and columns are called **marginal totals**, given by the last row (marginal totals for Weight) and the last column (marginal totals for Age). The marginal totals are a rough way to estimate the **marginal probability** of the attribute. For example,

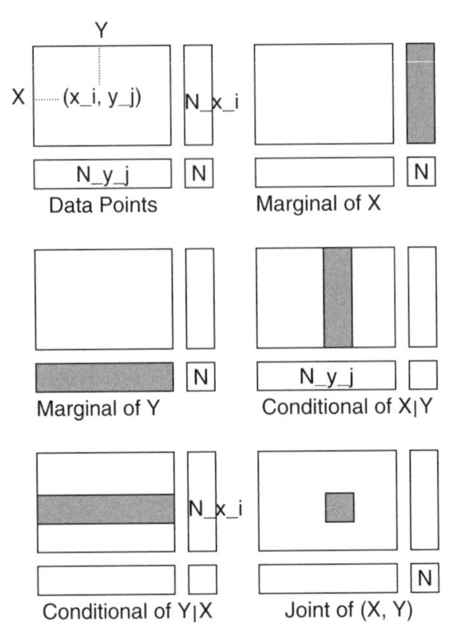

Figure 2.16: Probability distributions.

$$P(A = a) = \sum_{w} P(A = a, W = w) \tag{2.37}$$

is the marginal probability that age A has the value a, obtained by summing over all possible values of weight W. Estimates of the marginal probability of an attribute (say A) are given by

$$\hat{P}(A \in (10, 20)) = \frac{21}{55}. \tag{2.38}$$

The $\hat{P}$ notation is to indicate that the proportion is an estimate of the probability.

Suppose, we wish to know the behavior of the age attribute among organisms whose weight is restricted to the interval 5–10. Then:

$$\hat{P}(A \in (10, 20) \mid W \in (5, 10)) = \frac{10}{23}. \tag{2.39}$$

This is the proportion of organisms whose age falls in the interval [10,20], given that their weight lies in the interval [5,10]. The proportion is an estimate of the **conditional probability** of "age given weight". Note that this is different from the **joint probability** that "A lies in [10,20] and W in [5,10]" which is given by

$$\hat{P}(A \in (10, 20) \cap W \in (5, 10)) = \frac{10}{55}. \tag{2.40}$$

The relationship between the marginal, conditional and joint probabilities is given by

$$P(A \in I_1 \mid W \in I_2) = \frac{P(A \in I_1 \cap W \in I_2)}{P(W \in I_2)} \tag{2.41}$$

for all intervals I_1 and I_2. This property can be verified easily from Table 2.3.

Frequency tables are the intuition behind data summaries like the Empirical Cumulative Distribution Function and histograms that we discuss next. Such summaries are the next step in EDM and capture inter-relationships among attributes and variations over the entire attribute space. Contingency tables used to test the independence of attributes are based on two-way (or multiway) frequency tables. Similarly mutual information computation requires building a two-way frequency table. Fractal dimension computation too has a frequency table flavor to it.

2.6.2 Empirical Cumulative Distribution Function (ECDF)

The **Cumulative Distribution Function** (**CDF**) keeps track of the proportion of points that fall below any given value of an attribute. The CDF at $X = x$, for any $x \in S$, the support of X, is defined to be

$$F(x) = P(X \leq x). \tag{2.42}$$

We can think of the CDF as an inverse quantile function. That is, for a given probability p, the quantile function associates a value of the attribute $\xi(p)$ such that the probability that a data point has an attribute value less than $\xi(p)$ is p. The CDF does exactly the opposite, namely, associate a probability $P(X \leq x) = F(x)$ with every attribute value x. For example, the proportion of organisms whose weight is less than 10, from Table 2.3 is

$$\widehat{F_W}(10) = \hat{P}(W \leq 10) = \frac{36}{55},$$

where the hat notation indicates estimates and the subscript W references the attribute (weight) that the CDF F pertains to. Clearly, the proportion of points below $W = t$ increases as t increases. That is, there are more organisms whose weight is less than 20 than those that are less than 10, since all of the latter are included in the former category. (There will be at least as many, if there is no single organism with weight between 10 and 20.) Therefore, the CDF is non-decreasing.

The Cumulative Distribution Function can be estimated from the data in a very simple manner. The estimate is called the Empirical Cumulative Distribution Function.

$$\widehat{F(x)} = \frac{\textit{number of points} \leq x}{\textit{total number of points}} \tag{2.43}$$

$$= \frac{\sum_{i=1}^{n} I_i(x)}{N}, \tag{2.44}$$

where

$$I_i(x) = 1, \quad X_i \leq x \tag{2.45}$$

$$= 0, \quad X_i \leq x, \tag{2.46}$$

where X_i is the ith sample point. Variables that take values 0 or 1 based upon some condition (such as $I_i(x)$ above) are called **indicator variables**. N is the total number of data points. It is a step function since its value changes only at values of X observed in the sample. These change points are the ordered statistics of the data set.

The ECDF is a very powerful EDM tool. Steeper sections of the ECDF correspond to regions where we are more likely to find data points. For example, weights between 10 and 20 occur more frequently than weights between 80 and 90 in any sample. One can read off the median by drawing a line from the Y-axis at $p = 0.5$ to intersect the ECDF and drawing a line from there to intersect the X-axis. Other quantiles can be read off in a similar fashion. We can compare two groups (e.g. Snarks and Gryphons) by overlaying the ECDFs of the attribute. A visual inspection may quickly tell us that they are different. If they look similar, we can perform more rigorous tests.

Another EDM application of the ECDF is benchmarking against a known distribution. For example, if we suspect the distribution of age to be of a particular type, say exponential, we can overlay the ECDFs of the two and perform a quick visual check, followed by more rigorous tests if needed.

The ECDF itself is a random quantity since it is computed from the sample, therefore subject to sampling variation. The pointwise confidence intervals for the ECDF can be obtained by computing the confidence intervals for the corresponding quantiles. Packages like SAS automatically compute and plot

Table 2.4: Histogram: age by species.

Age	Species			Total
	Snark	Gryphon	Unicorn	
0–10	10	20	50	80
10–20	20	50	60	130
20–30	30	20	50	100
30–40	20	10	40	70
40–50	10	5	10	25
50+	5	10	10	25

these intervals. Simultaneous confidence bounds (as opposed to pointwise) for the entire ECDF can be computed as well.

The ECDF has nice EDM properties, namely, easy to compute, easily aggregated across many data sets or data sections, easy to visualize. It has good statistical properties as well (unbiased, consistent).

2.6.3 Univariate Histograms

Like the ECDF, a **histogram** is a representation of the concentration of data points over the attribute support. One of the frequently asked questions in EDM is, "How many data points lie in a certain interval?". For example, "How many people make more than 10 long distance phone calls a month?" More complex EDM questions need multivariate histograms, which we discuss in later sections.

Construction of a histogram involves two major steps.

1. Divide the support of the attribute into intervals or bins
2. Count the data points that fall into these bins

Consider the first row of data in Table 2.4, an example of a histogram of the ages of the organisms in the ecosystem. The first column represents the range of ages, between 0 and 10 units. The second column is the number of Snarks that are between ages 0 and 10. The last column represents the total for all the three species for that age group. Figure 2.17 shows a histogram of the above data.

Figure 2.17 is actually four histograms in one—one for each species, denoted by different shades of gray plus an overall histogram. The three together add up to fourth the overall histogram of age for the entire ecosystem. Note that in order to construct this summary we divide the range of the age variable into intervals and make no further distinction between points within this age interval. This is an important step in data reduction. The choice of the intervals clearly affects the histogram. There is extensive literature on

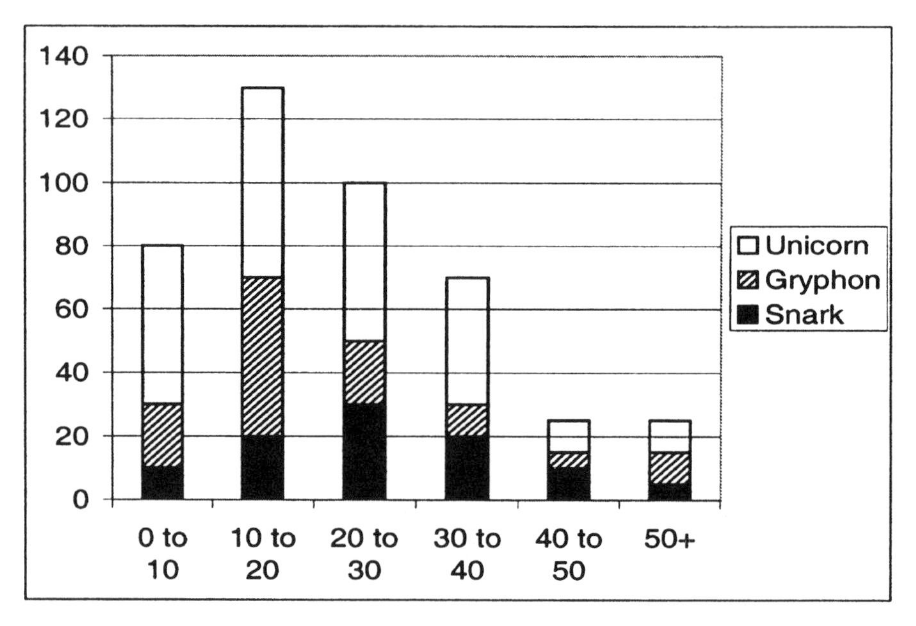

Figure 2.17: Three-in-one equi-spaced histogram of age.

the optimal choice of the histogram intervals. We will discuss two types of histograms in this section, the equi-spaced histograms and equi-depth histograms.

Types of Histograms

The example above is an **equi-spaced histogram**. We have divided the attribute age into intervals or bins of equal length. The last bin is an exception, it contains everything that spills over from the other bins. The advantage of this scheme is that the bins can be defined (in principle) a priori and independently of the data. The construction of an equi-spaced histogram requires only one pass over the data and is very quick. In addition, bins can be collapsed to create coarser equi-spaced histograms. Equi-spaced histograms make it easy to compare the distribution of points in different regions as well as compare histograms of different data.

An alternative approach is based upon having equal proportion of points (or mass) in each bin. Such a construction leads to an **equi-depth histogram** whose bin boundaries are quantiles.

Reconstructing Information from a Histogram

In order to retrieve information from a histograms, certain assumptions have to be made, since we can no longer distinguish between the points that are in the same bin. If there are 25 observations in the bin 40–50, we can use one of the following simple assumptions:

1. The observations are evenly (uniformly) distributed over the interval 40–50. The number of observations is proportional to the length of the interval and does not depend on the actual values of the data points. So the segment 40–42 (one-fifth the length of the interval 40–50) would contain one-fifth the number of observations in 40–50, namely, 5.
2. All the observations are centered at the mid-point, namely, 45. That is, there are 25 observations all of which have the value 45.
3. More complicated assumptions require the retention of additional information. For example, we could keep the mean of the 25 observations that fall in this bin while creating the histogram, and assume that all the 25 values have the same value as the mean.
4. Distributional assumptions about the allocation of the points in the bins. For example, instead of a uniform distribution, they have a normal distribution around the mean. This requires storing information other than just the bins and bin counts.
5. Using splines to approximate the distribution of points inside the bins.

On the basis of such assumptions we can now recover other information from the histogram. Under the second assumption, we can compute the following:

- The mean age is given by (for species A)

$$\overline{X_A} = \frac{5*10+15*20+25*30+35*20+45*10+55*5}{95} = \frac{2525}{95}. \quad (2.47)$$

- Probability that age (of any individual from species A) is less than 30 is given by

$$P(Age(A) < 30) = \frac{10+20+30}{95}. \quad (2.48)$$

Note that for the last bin we are assuming that the mass is concentrated at 55.

2.6.4 Annotated Bibliography

Multivariate distributions and the relationship between conditional, joint and marginal distributions are described in most texts on probability and statistics. See [105], as an example. Well-known nonparametric techniques for exploring and understanding multivariate distributions include clustering [69], some pattern recognition techniques [38], hierarchical networks [99] and neural networks [9].

Estimating univariate distributions through histograms and other estimates like the ECDF has been covered in many books. See [63] for and advanced discussion of the ECDF, and its properties. A discussion of **simultaneous confidence bounds**, which are stronger than pointwise confidence bounds, is available in [20]. See [115] for detailed discussions of histograms, their construction, density estimation and other related concepts. The use of splines for recovering information from histograms is described in [91]. Given the importance of the choice of the bins, there has been a lot of interest in determining optimal bin selection. An important approach is discussed in [67].

2.7 EDM IN HIGHER DIMENSIONS

An important challenge in large-scale data analysis is the "curse of dimensionality". As the number of attributes (also known as dimensions) increases, analytical techniques slow down, often in exponential or polynomial proportion to the number of dimensions. Techniques based on computational geometry seldom go beyond three or four dimensions. In general, nice properties that hold in lower dimensions disappear in higher dimensions. Visual techniques all but fail after five dimensions, becoming increasingly complex with color coding and animation. Ordering data in two or more dimensions requires defining special transforms to induce a one dimensional order. In addition, there is "space warp", that is, data points tend to get packed at the boundaries leaving large tracts of empty space (see [115]). Small perturbations in the data can cause large shifts in the results of the analysis.

We first discuss below a natural extension of the univariate histogram by using a multivariate rectilinear grid. A discussion of scalability will follow in Chapter 3.

2.8 RECTILINEAR HISTOGRAMS

Table 2.3 which we used in the frequency table example, is a rudimentary two-way histogram. Each combination of ranges of the two attributes is a bin. For example, constraining the age to lie between 10 and 15 AND the weight to lie between 15 and 18 would correspond to a rectangular region, a "bin" in two dimensions. If all of the attributes are numeric, the bins of a rectilinear histogram are specified as

$$X_1 \in [x_{1l}, x_{1u}] \cap X_2 \in [x_{2l}, x_{2u}] \ldots \cap X_d \in [x_{dl}, x_{du}], \tag{2.49}$$

where each attribute X_i has to lie in a specified interval $[x_{il}, x_{iu}]$ and l and u in the subscripts denote lower and upper bounds of the interval. One or more of the attributes of the histogram can be categorical. For example, Table 2.4 is a two-way histogram with a categorical attribute. The categorical

Table 2.5: Histogram binning scheme.

Original			Binned		
Age	Weight	Volume	Age	Weight	Volume
8	12	9	0	0	0
6	11	12	0	0	1
7	17	8	0	1	0
8	16	5	0	1	0
12	13	9	1	0	0
13	13	9	1	0	0
15	12	7	1	0	0
12	13	17	1	0	1
15	13	16	1	0	1
14	12	14	1	0	1
18	14	16	1	0	1
18	20	9	1	1	0
19	19	9	1	1	0
12	18	9	1	1	0
15	16	8	1	1	0
22	19	9	1	1	0
22	25	20	1	1	1
18	18	15	1	1	1
16	15	22	1	1	1

attributes can also be "binned" into coarser subdivisions, as is discussed in Section 3.2.

Let us consider the ecosystem example. There are three numerical attributes—age, weight, and volume. We construct an artificial example by selecting two bins for each attribute 1 (high values) and 0 (low values) by choosing some cut off values (e.g., age cut-off = 10, weight cut-off = 15, volume cut-off = 10). Table 2.5 shows the mapping from the data values to the bins, which is summarized into the multivariate histogram shown in Table 2.6.

The last column of Table 2.6 represents the number of organisms with that particular combination of the three attributes. We can see that the histogram is a compact representation of the raw data. There is a loss of information caused by binning the variables age, weight and volume into two bins each 0 and 1. We can increase the granularity and use three bins for each attribute (−1 (low), 0 (medium), 1 (high)) resulting in a total of $3^3 = 27$ combinations as opposed to the current $2^3 = 8$. As the number of attributes increases, the rectilinear histograms become unmanageably large. For example, a histogram based on merely 6 attributes and 10 bins for each attribute results in

$$10^6, \tag{2.50}$$

a million bins!

Table 2.6: Multivariate histogram on age, weight, and volume.

Age	Weight	Volume	Count
0	0	0	1
0	0	1	1
0	1	0	2
0	1	1	0
1	0	0	3
1	0	1	4
1	1	0	5
1	1	1	3

The problem is mitigated somewhat by the use of *data cubes*, which permit aggregation based on the levels of categorical attributes. Datacubes are discussed in the next chapter. The *DataSphere* partitioning scheme is a nonlinear technique that allows scalable partitioning, resulting in a number of bins that is related *linearly* to the number of dimensions, as opposed to exponentially. In the next chapter, we explore data cubes, DataSpheres, potential partitioning schemes based on depth and their advantage over context specific partitions induced by models like classifiers.

2.9 DEPTH AND MULTIVARIATE BINNING

Data depth denotes how deeply embedded a data point is in the data cloud. Depth is relevant to EDM and DQ for several reasons:

- Data depth is important for many EDM tasks such as understanding the distribution of data points in space. Such information is useful for identifying outliers, for finding which points are "close" to each other, and so on.
- We can use depth based bins for building a skeleton of the data through multivariate binning. The skeleton can then be used for data aggregation and summarization, including counts, sums, cross products and sums of squares. The aggregates, in turn, can be used for data publishing.
- In higher dimensions, parametric analysis based on assumptions about the underlying distribution becomes harder due to the mathematical complexity of the functional forms *f*. We can used depth to partition the attribute space and use simple nonparametric methods to approximate complex data structure.

In this chapter, we will introduce depth, to lay the ground work for the next chapter on partitioning and nonlinear EDM.

2.9.1 Data Depth

Picture the following: Apples are arranged in a display on a barrel in a grocery store. The barrel can be accessed from all sides. In order to get to the one really good apple which the grocer has malevolently hidden in the center of the display, we would need to carefully remove the outer apples. The good apple is deep inside the crowd of apples. If we pretend the apples are data points in three-dimensional space, data depth is a measure of how deeply entrenched a point is with respect to the other data points. We motivate below some intuitive notions of depth.

In the case of a single attribute, the data points can be plotted on the real line. As we traverse the real line from either of the extreme values (the Min or the Max of the data points) we will pass through the sample points. The more points we need to cross in order to get to a given point, the deeper it is inside the data set. Clearly the median is the deepest point.

Next consider a data set with two attributes. We can imagine a straight line sweeping through (without rotation) the data cloud to reach a given data point. The line that passes through the smallest number of points will determine the depth of the point, given by the number of points it has to pass through. Alternately, we can think of the data points organized into layers, as in an onion. All the outermost points lie on a layer that can be peeled away and the next set of points on another layer and so on recursively until we reach the very center (core of the onion). In such a formulation, a rough measure of depth would be the number of layers that need to peeled off in order to get to the point of interest. Yet another intuition for depth is based on the construction of triangles. If we consider the bivariate case as an example, we can construct triangles by connecting all possible sets of three data points that do not lie in a straight line (noncollinear). (There will be $\binom{n}{3}$ such possibilities). Points on the periphery will be contained in fewer triangles than the points towards the interior of the cloud. The number of the triangles that contain the point of interest is an indication of how deeply embedded the point is in the data cloud. Thus there are many ways of defining depth in multivariate space. Several formal definitions of multivariate depth exist. We will discuss some well-known definitions below. Note that they are all applicable to a general d-dimensional space, in principle. However, the expense of computing the depth might be prohibitive for some definitions when the number of data points and/or the number of attributes increases.

- **Mahalanobis Depth**: The Mahalanobis depth of a point x from a multivariate density with mean μ and covariance matrix Σ is given by

$$d(x) = (x - \mu)'(\Sigma^{-1})(x - \mu). \tag{2.51}$$

  This is also called the statistical distance. The Mahalanobis distance requires the existence of second order moments.

- **Half-plane Depth (Tukey Depth)**: Consider various planes that pass through a point x in multidimensional space. Each plane partitions the space into two parts. With each halfspace, we can associate a likelihood of finding data points. Empirically this would be the proportion of data points contained in that halfspace. The smallest such probability over all possible halfspaces associated with x is its Tukey depth.
- **The Convex Hull Peeling Depth**: If the data set consists of n points, construct a convex hull containing all the n points. We can think of a convex hull as the smallest convex polygon that contains the outermost points of the data cloud. If the data were scattered in a sphere, we can think of convex hulls as the layers of an onion. We start by peeling off the outermost convex hull (called a layer of depth 0), thus removing all the points lying on it. Next, we construct the convex hull of the remaining points. This is a convex layer of depth 1, and so on. The depth of a point is the depth of the convex layer it lies on.
- **Simplicial Depth**: We give here an informal definition. If x is a point in d dimensional space, consider the convex hulls of all subsets formed by a specified number $(d + 1)$ of points. When these are well defined (the points do not lie on a straight line), such a convex hull is called a simplex. The simplicial depth is the probability that a point x will lie in a simplex. The sample version is the proportion of simplices that contain x. In Figure 2.18, the point represented by the big dot, is enclosed in many simplices (triangles in this case) as compared to the point represented by the square. Hence the big dot has greater depth that than the square.
- Other concepts of depth include the likelihood depth based on the density at x. Another related concept is **regression depth**, which measures the depth of a regression plane with respect to the data. We relate this concept to goodness-of-fit and outlier finding in Section 5.2.6.

With the exception of Mahalanobis depth, the depth metrics are computationally intensive. However, in situations where it is feasible, computing other depth metrics offer insights about the joint distribution (half-plane depth, convex hull peeling depth, likelihood based depth) and its structure that are valuable in their own right. (See Section 2.9.2 below.)

2.9.2 Aside: Depth-Related Topics

Depth Median

Let us revisit the notion of multivariate center, using depth. In the univariate case the median can be defined to be a single data point. In higher dimensions, it is more common to define a median region, the region containing data points with the maximum depth. For example, the median region would be the region contained in the innermost convex hull obtained by peeling. We have seen that in the univariate case the median is a robust alternative to the mean. The

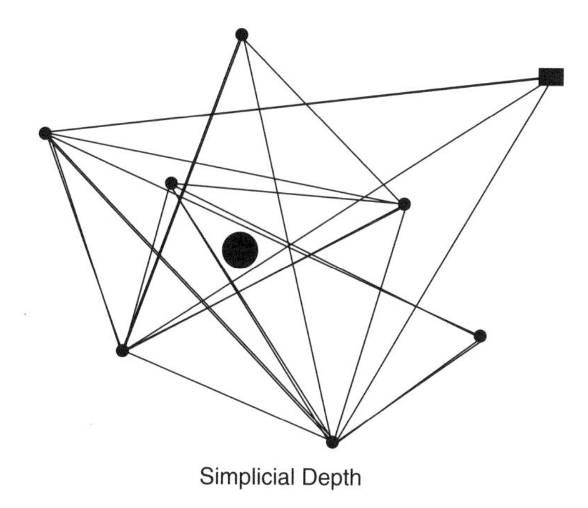

Figure 2.18: The big dot has greater simplicial depth than the square.

robustness of a center is important, especially when it is the parameter that defines a partitioning scheme for data reduction as in DataSpheres, a nonlinear partitioning scheme we discuss in the next chapter.

Depth Contours

Points that have the same depth lie on the same depth contour or level curve. Depth contours differ from probability contours. The latter represent areas of equal density, while depth contours pertain to relative location. Depth contours are important for:

- Outlier detection—Usually peripheral points or outliers are different from the majority of the data points. Their extreme nature makes them (a) interesting as in identifying high-usage customers or data glitches, or (b) a nuisance as in skewing averages and typical behavior. In either of these situations it is useful to identify them and separate them from the rest of the data set. The outermost depth contour (or layer of depth 0) contains multivariate outliers. We will return to outlier detection in a Section 5.2.3.
- Change in density—The algorithm that computes the depth contours keeps track of the number of points that fall inside each depth contour. Therefore we can compute the change in the density as we peel off consecutive layers.
- Finally, depth contours can be used to construct a scalable partition of multivariate data space for data aggregation. We can define a **depth equivalence** class, or **de-class** as a set of points that fall within a certain depth range. Each de-class could be a class in the partition.

2.9.3 Annotated Bibliography

See [83] for an overview of multivariate depth and related concepts. A discussion of simplicial depth can be found in [82]. See [124], Tukey's paper on stochastically equivalent blocks for a description of *de*-classes.

For an introduction to convex hulls and their computation see [101]. See [89] and [121] for computation of location depth and depth contours. See [111] and [112] for more on regression depth.

2.10 CONCLUSION

In this chapter we introduced statistical summaries that play an important role in EDM. We discussed the characteristics of good EDM summaries. Some of the summaries can be used for detecting glitches in massive data and screening them for data quality purposes. We discussed both model-driven summaries as well as data-driven summaries like quantiles, histograms, ECDF and others. We used data depth to generalize the concept of quantiles to higher dimensions and set the stage for scalable multivariate space partitions or bins. Such partitions will be used extensively in the next chapter to find fast, simple EDM models for capturing nonlinear relationships among attributes.

CHAPTER 3

Partitions and Piecewise Models

3.1 DIVIDE AND CONQUER

Experience tells us that a difficult problem can be tackled better if it is broken down into smaller pieces. For example, exploring a new city can be done by dividing it into regions and exploring each piece separately and systematically until all the regions are covered. Similarly, a large intractable data set can be divided into smaller pieces for easier exploration and analysis. Furthermore, large data sets are usually heterogeneous and consist of a mixture of probability laws. We are interested in unraveling the mixture and identifying each component to the greatest extent possible. For example, clustering techniques divide the data into clusters of like data points (married with kids, single and employed, retired but less than 70 years of age). Such groupings are very popular in advertising and marketing campaigns to target consumers. Classification techniques (machine learning, neural networks, logistic regression) divide the data into "classes" which constitute meaningful pieces of the data set (likely to switch phone companies, unlikely to switch phone companies). There are more general ways of dividing the data set as well.

In this chapter, we discuss methods for dividing up the data into manageable pieces for the purpose of (a) data summarization, (b) data cleaning, and (c) scaling up analyses to massive data through piecewise models built using EDM summaries. In the rest of this section, we motivate partitions and partition based analysis. In Section 3.2, we discuss linear partitions. In Section 3.3 we focus on nonlinear partitions. DataSpheres, a particular type of nonlinear partitions are discussed in Section 3.4. In Section 3.5, we describe set comparison techniques based on EDM summaries of partitions for detecting changes in data sets, whether caused by genuine shifts in distributions or by data glitches. Section 3.6 describes the process of approximating complex structure in data using a collection of simpler models. In Sections 3.7 and 3.8,

Exploratory Data Mining and Data Cleaning, by Tamraparni Dasu and Theodore Johnson
ISBN: 0-471-26851-8

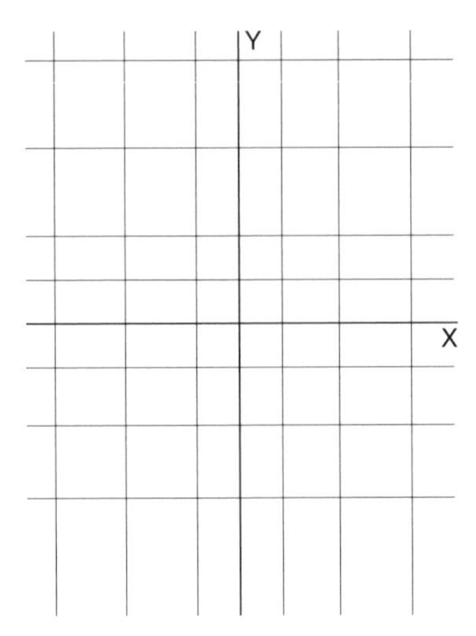

Figure 3.1: Rectilinear boundaries.

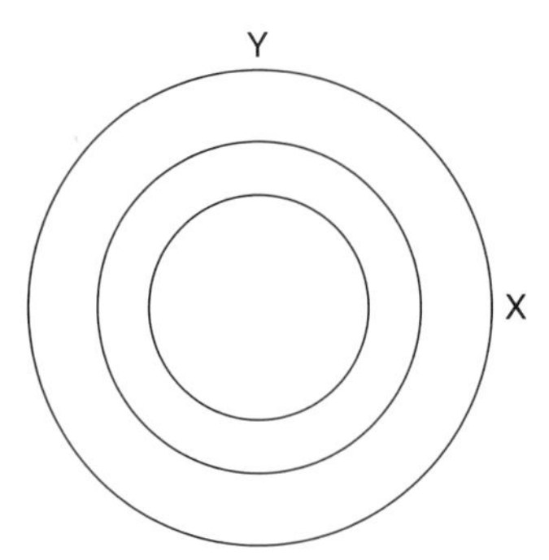

Figure 3.2: Curved distance partitions: $X^2 + Y^2 = C^2$.

we give instances of such approximations, namely piecewise linear regression and one-pass classification. We conclude with a summary in Section 3.9.

The process of dividing the data set into non-overlapping parts that account for the entire data set is called **partitioning** and each resulting piece of the data set is called a **class** of the partition. Figures 3.1 and 3.2 show examples of partitions with different types of boundaries.

3.1.1 Why Do We Need Partitions?

Partitions serve many useful purposes.

- **Simplify:** A partition provides a simplified, high-level, schematic version of the data set, similar to the human skeleton. It makes it possible to identify different portions of the data at a high level, study their inter-relationships (the back bone is connected to the hip bone . . .) and identify interesting hypotheses (the skull has a hairline fracture). The hypotheses can be investigated in greater detail by looking at that portion of the data (the head region, skin lacerations, muscle, blood vessels) in greater detail.

- **Summarize:** In the context of EDM, the classes of the partition can be used as a binning scheme to collect aggregates. Histograms are examples where we used the bins to collect counts of the number of data points that lie in a bin. We can also collect other aggregates such as sums, sums of squares, min, max, a sample, and so on, which can be used for data exploration, visualization, exploratory model fitting, data publishing and for data cleaning purposes. The process of creating a smaller representation of the data using summaries is called **data reduction**.
- **Reduce Variability and Data Set Size:** Data partitions help us treat the data at a level of granularity that suits our constraints of resources and time. Computing aggregates of the whole data without any partitioning (the coarsest level of analysis) hides most of the interesting variations and features of the data. Treating every data point as a class of its own (the finest level of granularity) results in too much noise, so that it is not possible to capture general patterns that are applicable to large portions of the data. Besides, it might be computationally impossible to treat data at its raw granularity in the case of massive data. *The optimal choice of granularity for a partition lies between these two extremes, where there is sufficient reduction in variability to dampen noise but enough detail to preserve interesting local structure in the data.*

3.1.2 Dividing Data

The data set cannot be apportioned arbitrarily. It has to be done in such a fashion that interesting structure in the data is preserved and can be inferred from the smaller pieces of the partition. For example, if we were to randomly pick fragments of the human body and jumble them up indiscriminately into classes (combine the thigh bone, one vertebra and a fragment of the skull into one class) we will not gain valuable insights. Every class will be a garbled mess. Therefore, it is important to have some notion of contiguity and similarity while creating a partition. However, the constraints should not be so strong that we cannot use the partition and the EDM summaries in a general context.

Partition *induced* by context-specific techniques (clustering, classification) (Figure 3.3) are **parameterized** by the technique, just as parametric methods are based on assumptions about distributions and inter-relationships between attributes. Therefore, while these induced partitions are attractive because they result in a handful of meaningful classes, they are less useful in more general contexts of data summarization and data publishing. During the data exploration phase of an analysis, we cannot be sure that the models that induce

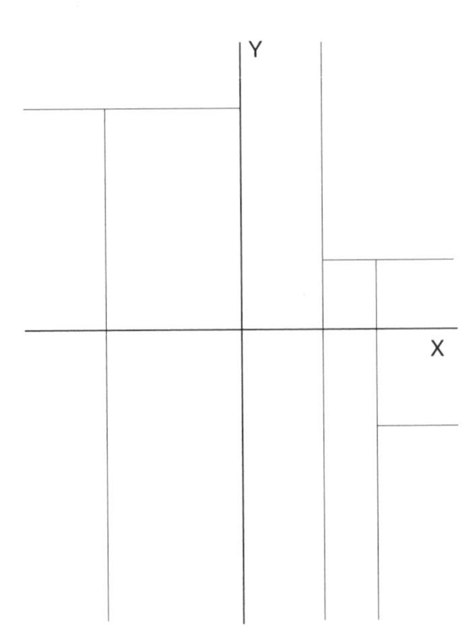

Figure 3.3: Partition induced by decision trees.

the partition are appropriate for the data. When they are, the models tend to be closely tailored to the specific data set at hand, so that they do not generalize well to other data sets. Therefore, during data exploration it is better to choose an exploratory partitioning scheme rather than a partition induced by a model designed for a specific objective, based on several tunable parameters and assumptions. Finally, some context-specific partitioning techniques are computationally expensive, again making them inappropriate for data exploration.

The task of creating multipurpose partitions is relatively straightforward in the univariate case (e.g. bin using quantiles of the attribute) but it gets complicated in higher dimensions. As mentioned earlier, choosing bins componentwise and considering all combinations results in a rapid (exponential) increase in the number of classes of the partition. However, such partitions are simple to interpret (age between 50 and 60, income between \$50 K and \$100 K) and therefore attractive. Furthermore, such schemes fit nicely into a data management scheme called *data cubes*.

In this chapter, we will focus primarily on two nonparametric partitions—(1) the axis aligned partitions mentioned above called *data cubes*, and (2) *DataSphere*, a nonlinear scalable partitioning strategy. (Induced partitions should be studied in the context of the technique that generates them, e.g., clustering. They are not a topic for this book.) We will use the nonparametric partitions to create a set of EDM summaries.

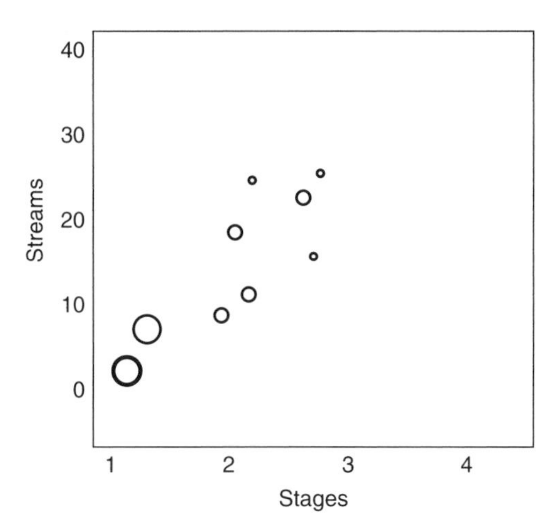

Figure 3.4: Visualizing cluster sizes.

3.1.3 Applications of Partition-Based EDM Summaries

We outline below the analysis applications of partition-based EDM summaries that we discuss in this chapter.

- **Visualization** is an effective and intuitive part of EDM. However, due to the limitation of display media as well as the human eye, visualizing data in more than 4–5 dimensions (aided by, e.g., animation, color coding) is difficult. Complicated projection tours and other techniques require considerable expertise and often break down for large data sets. In this context, partitions and partition-based EDM summaries play an important role in reducing the dimensionality as well as the size of the data, facilitating visualization. There are examples of partition-based visualization throughout the chapter as well as the rest of the book. We will briefly mention a few below.

  First, consider an ordinary histogram or bar chart. It facilitates visualization by plotting aggregates (bin counts) rather than the raw data. Plotting the raw data would result in an indecipherable mass. Next consider the cluster size plots in Figure 3.4. The clusters are computed in six dimensions based on around 17,000 data points, each of which represents a provisioning process. The size of the circles is proportional to the size of the cluster. A circle is centered at the cluster means of two of the attributes *streams* and *stages*, which are measures of complexity of the process being depicted. The plot helps us understand the relative sizes of the clusters (partition classes) as well as their relationship to two of the clustering attributes.

Another kind of visualization is facilitated by the *DataSphere (DS)* partition discussed below, where a multi-dimensional space can be represented in two dimensions that uniquely determine the class of a partition. We defer a discussion to the section on set comparison where the *DS* partition based **bead plots** (see Section 3.5.4) are used for illustrating the dispersion of points in high-dimensional space.

- **Set Comparison and Detection of Data Glitches**—An important use of the EDM summaries is the comparison of two or more massive data sets. We can rapidly identify subgroups that are different (elementary school students in New Jersey are statistically different in their academic performance from those in California) as well as detect sections of the data that look suspiciously different from a well-known standard. We will discuss set comparison as a powerful data cleaning technique in the next chapter.
- **Interactive Model Fitting**—The EDM summaries can also be used to learn the structure of the data by rapidly building approximate models that can point the way to further, more sophisticated models (e.g., nonlinear regression will work, parametric failure models not appropriate since assumptions are not valid, etc.). Such *exploratory model fitting* is an important application of EDM summaries because randomly trying different models from a suite of data mining methods is very expensive. In this chapter, we will give two examples of piecewise model approximation with EDM summaries.

3.2 AXIS-ALIGNED PARTITIONS AND DATA CUBES

During the 1990s, the database community (both research and industrial) has developed technologies (e.g., *On Line Analytical Processing (OLAP)*, star schemas, data cubes, etc.) to enable users to explore large data sets. Many of the ideas derive from earlier work on statistical databases, but they have been refined to allow them to better apply to large and complex **data warehouses** (databases specifically for long-term storage and analysis of data sets). This material is well covered in other books, so we provide only a summary here.

A data warehouse will contain one or more **fact tables**, which contain the data to be analyzed. The fields of a fact table can be either **dimensional** attributes, which describe the entity represented by an entry in the fact table, or **measure** attributes, which are measurements of the entity. (In Statistical terminology, the values of the dimension attributes are called the *levels* of a categorical variable.) For an example, let us recall the data set from Chapter 2 that describes the Himalayan ecosystem. The fields of this fact table are:

species, age, weight, volume.

The Species field describes the measured animal, hence it is a dimension. Age, Weight, and Volume are measures.

Our data warehouse is likely to contain supplementary information about each of the species—for example, their myth of origin, fur color, relation to Lewis Carroll, and so on. These fields describe the species rather than the individual, so good database design and common sense dictate that we store the information about species in a separate **dimension table**:

AnimalSize(SpeciesID, Age, Weight, Volume)
SpeciesInfo(SpeciesID, Name, Myth, FurColor, Carroll)

Since we are dealing with multiple tables, we need to assign each table a name, for example, AnimalSize. Entries in the two tables are related using the SpeciesID field. Given an entry in the AnimalSize table, we can find the name of the species of the measured animal by looking for the entry in the SpeciesInfo table with the same SpeciesID, and using the value of the Name field. This process is referred to as the **join** of the two tables, and is a basic activity in relational databases.

When these measurements are collected, there is often a great deal of additional descriptive information available. For example, the scientist might record the year in which he makes the measurement, the location of the measurement, and so on. Therefore our fact table will look like the following:

AnimalSize(SpeciesID, TimeID, LocationID, Age, Weight, Volume)
SpeciesInfo(SpeciesID, Name, Myth, FurColor, Carroll)
MeasurementTime(TimeID, Year, Month, Day, Hour)
Location(LocationID, MountainName, Face, Height)

When the tables and their relations are visualized, the fact table (AnimalSize) will be in the middle with the dimension tables radiating out from it and linked by their join relationship. Hence this type of database organization is often called a **star schema**. **On-line Analytical Processing (OLAP)** refers to the activity of analyzing data sets in a data warehouse, especially data sets stored using star schemas. One of the main purposes of OLAP systems is to allow users to easily and quickly issue data exploration queries, which usually partition the fact table using the dimension fields and summarize one or more of the measure fields. An example query is "For each year from 1990 to 2002, report the average weight of Gryphons on the North face of Mt. Everest."

Answering these questions on a very large data set can be painfully slow. One way to speed up the processing is to precompute some of the summaries, and use them to compute the answer. Taking this idea to its logical conclusion, a **data cube** is the collection of all possible summaries. Since the answer to all summarization questions have already been computed, we can provide the user with an answer by looking it up rather than by computing it (which is likely to be much faster). For example, suppose that the queries involve

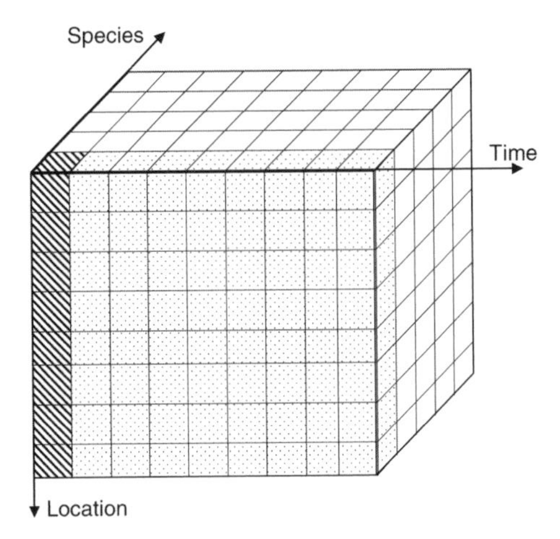

Figure 3.5: A three dimensional data cube.

average weights and counts of the number of animals. Then the data cube for answering these questions is:

AnimalSummary1(SpeciesID, TimeID, LocationID, Avg(Weight), Count)
AnimalSummary2(TimeID, LocationID, Avg(Weight), Count)
AnimalSummary3(SpeciesID, LocationID, Avg(Weight), Count)
AnimalSummary4(SpeciesID, TimeID, Avg(Weight), Count)
AnimalSummary5(SpeciesID, Avg(Weight), Count)
AnimalSummary6(LocationID, Avg(Weight), Count)
AnimalSummary7(TimeID, Avg(Weight), Count)
AnimalSummary8(Avg(Weight), Count)

A data cube is useful not only to speed up answers to OLAP queries, but also to help understand the data set. By comparing the tables *Animal Summary3(SpeciesID, LocationID, Avg(Weight), Count)* and *Animal Summary6(LocationID, Avg(Weight), Count)*, we can quickly see which species are dominant in a location and whether their weight is significantly different from the average. The name "data cube" is derived from visualizing three-dimensional data and each of the marginal summaries, for example in Figure 3.5. The lightly shaded area corresponds to AnimalSummary2, while the darkly shaded region corresponds to AnimalSummary6.

The dimensions of a data cube can have additional structure. For example, the MeasurementTime table contains a **hierarchy** of Years, Months in the Years, Days in the Months, and Hours in the Days. Therefore, we can select the granularity of summarization by selecting a position in the Measurement-Time hierarchy (e.g., average over months vs. average over hours). Furthermore, a dimension can contain multiple orthogonal hierarchies. For example, a position in the Location table can be further refined by specifying the Face of the mountain, the Height on the mountain, or both. These two subdimensions might have dimension tables of their own, producing a **snowflake** hierarchy.

A variety of software packages provide dimensional "data cube" summarization. For example, modern spreadsheet packages will compute two-dimensional data cubes, called **pivot tables**, but the dimensions can be hierarchical. Most modern database systems provide extensive OLAP and dimensional analysis facilities.

These types of powerful data summarization products are a useful and readily available tool for data exploration. The use of hierarchical dimensions is particularly valuable because it reduces the apparent dimensionality of the data set. For example, our Himalayan animal measurement data set would at first appear to have eight dimensions (species name, year, month, day, hour, mountain, face, height), but they can be arranged into a more manageable three dimensions.

As can be seen from the list of tables in a data cube, OLAP software allows the user to readily specify the level of detail to be presented. Each of the tables of a data cube represents a **slice** of the data set. The detail in the report can be reduced by **rolling up** on one or more dimensions. Conversely, the detail in the report can be increased by **drilling down** on one or more dimensions.

3.2.1 Annotated Bibliography

Data cubes were proposed by Gray et al. [54]. Since then, an extensive literature has been published. A survey of the literature can be found in [18]. Most of the commercial database engines provide OLAP support, we recommend reading their manuals.

3.3 NONLINEAR PARTITIONS

Nonlinear partitions are important in the context of reducing the dimensionality of a dataset and capturing the interaction between attributes. For example, if we can partition a data set into clusters, a data point can be identified by its cluster membership and perhaps a few characteristics (cluster center, cluster diameter) that define the cluster. We need not keep track of the individual attributes that define the data point. We will not discuss clustering in this book but merely allude to it since it is a potential space partitioning technique. Informally, clustering involves grouping "similar" data points together so that the points in the same cluster resemble each other more than points in any other cluster. Similarity is defined based on various metrics such as distance, density and other criteria. Such grouping into well-defined and well-separated clusters is much sought after, especially in consumer marketing studies. However, clustering requires a fair amount of domain expertise, experience with the algorithms and the tuning parameters, and might prove to be computationally expensive.

In the next section, we focus on a fast and nonparametric nonlinear partition which is called the **DataSphere (DS)** partition. This partitioning technique has been proposed for the rapid exploration and understanding of large, high-dimensional data sets. The construction of the DataSphere is akin to making a mold for the data. The mold can then be used for a variety of purposes. We can test to see if other data sets "fit" our mold. If they fit snugly, we can infer that the data sets are similar. If not, we can identify the regions of the mold where the fit is not good. These concepts play an important role in the automatic detection of glitches in massive data sets. We formalize the notion of testing the fit of a DataSphere mold to a dataset using set comparison techniques described in Section 3.5 of this chapter.

3.3.1 Annotated Bibliography

For an overview of clustering see [69]. For a discussion of validity of clustering and the different ways of measuring goodness-of-fit of clusters, see [68]. Note that implementing the tests for validity of clusters with respect to the data could involve MCMC (Markov Chain Monte Carlo) simulations, making it time consuming. Such tests are often not implemented. For computationally feasible algorithms for massive data, see [51]. For clustering mixtures of Gaussians see [28].

DataSpheres were introduced in [29], and a more evolved description including the effect of different centers and scaling parameters on the boundaries of a DataSphere is presented in [72].

3.4 DATASPHERES (DS)

The *DS* method partitions the data based on two criteria for the numeric attributes, that of distance or depth, and direction. Each class in the partition is uniquely defined based on the combination of these two criteria. Using "data cube" terminology, they are two of the dimensions of the entries in our fact table, while the categorical attributes provide other dimensions.

To use a DataSphere, we compute EDM summaries for every class by calculating specialized aggregates based on the points that fall into the class. The DataSphere summaries, which have special properties, described in Section 3.4.3, are used as a basis for EDM and further analysis, including visualization. These summaries are usually provided by OLAP software packages. By treating DataSphere depth and direction as two of the dimensions in a data cube, OLAP software packages can be readily and profitably employed.

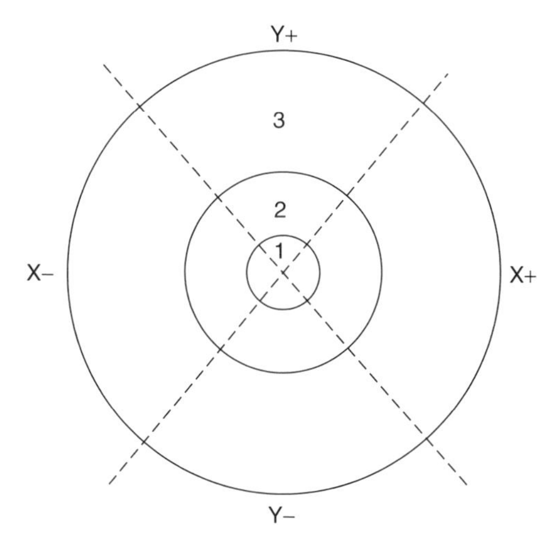

Figure 3.6: DataSphere partition in 2-D.

3.4.1 Layers

The first step in creating a DataSphere is defining **depth layers** using an appropriate subset of the numeric attributes. The attributes that are used to compute the layers are called **depth attributes**. The layers divide the data into concentric depth shells (like the layers of an onion), based on a suitably chosen measure of depth. If the data set is unfamiliar, all the numeric attributes should be treated as depth attributes. In the ecosystem example, weight, age and volume are potential depth attributes. The categorical ("dimension") attributes are also used to partition the data. In the context of statistical partitioning, such variables are called **cohort attributes**. For example, species is a cohort attribute. Every combination of cohort attributes defines a **subpopulation** of the data. For example, Snarks constitute a subpopulation.

For the *DS* computation, we use a center outward approach for depth, by choosing a **reference center** and computing the depth of every point relative to that center. Points that fall within the same depth range constitute a **depth layer**. This approach is attractive because it is computationally tractable and has a nice interpretation. From our discussion in the previous chapter, the inner depth layers (Figure 3.6, layers 1, 2) close to the center represent "typical" or representative data points. The outer depth layers correspond to outlying or abnormal data points, presumably from the tails of a multivariate distribution f. We have observed in practice that the technique will work despite the strange behavior of data in higher dimensions, where it tends to concentrate at the boundaries.

The depth layers can be computed as follows:

- Compute a *center* for the data cloud using the depth attributes. Practical choices include multivariate mean, multivariate trimmed mean and componentwise median.
- Center and rescale the depth attributes using the center computed above and using an appropriate measure of dispersion such as the standard deviation or interquartile range. It seems natural to pair the mean and the quadratic deviation measured by the standard deviation, and the median and the absolute deviation measured by the MAD, as per the discussion in section on typicality.

  A data point $X_i = (x_{i1}, x_{i2}, \ldots, x_{id})$ will be transformed to Y_i where

$$\begin{aligned} Y_i &= (y_{i1}, y_{i2}, \ldots, y_{id}) \\ &= \left(\frac{x_{i1} - \overline{x_1}}{\sigma_{x1}}, \ldots, \frac{x_{id} - \overline{x_d}}{\sigma_{xd}} \right), \end{aligned}$$

  where $\overline{x_j}$ and σ_{x_j} are the mean and standard deviation, respectively, of the jth component. We can replace the mean and standard deviation with other choices. Standardizing the data makes attributes free of measurement units and scales, making them comparable.
- Compute the distance d_i of every point from the center.

$$d_i = \sqrt{\sum_{j=1}^{d} y_{ij}^2}.$$

  We have used the Euclidean distance, but other choices such the Manhattan (L_1) distance can be used too. The choice of distance effects the DataSphere partition boundaries. Other depth boundaries (depth contours, convex hulls), while meaningful and data driven, are geometrically complex and difficult to compute.
- Sort the data points by distance d_i from the center and define the layer boundaries to be distance quantiles. That is, all data points whose distance lies between two consecutive distance (or depth) quantiles constitute a layer. The number of layers can be chosen at will. Extra layers can be collapsed into a coarser partition subsequently, if needed.

  Depth quantiles are computed efficiently using recent one-pass fast quantiling algorithms. Using the distance quantiles as layer boundaries ensures that there are roughly the same number of data points in each layer.

The center and the distance layer boundaries, called the **DataSphere parameters**, together define a unique DataSphere representation of the data. Note that in higher dimensions, sparseness and concentration of data points along boundaries might be an issue. In practice, however, we found the technique to be effective in quite high-dimensional data sets. Our center is merely a reference point and therefore meaningful even if the data is concentrated along boundaries. Furthermore, the distance layers give an indication of the dispersion. The inner layers are usually close together, with the distance between consecutive layers exploding as we get to the outer layers.

A partition of the data space based just on distance layers has no directional information. We do not know the contribution of each attribute to the distance, so that an observation that lies in layer n can be either due to extremely high values of attribute *Age* or extremely low values of attribute *Weight*. In order to incorporate directional information we refine the partition using *data pyramids*, described below.

3.4.2 Data Pyramids

A data point can be characterized by the attribute that influences the distance most, the one with the maximum deviation from the center (measured in standardized values). For example, we need to distinguish between outliers in *Age* apart from the outliers in *Weight*. We can do this (to an extent) using the concept of **pyramids**. Briefly, a d-dimensional set can be partitioned into $2d$ pyramids $P_{i\pm}$, $i = 1, \ldots, d$ whose apexes meet at the center of the data cloud. That is, for any point data point p

$$p \in P_{i+} \quad \text{if } |y_i| > |y_j|, y_i > 0 \quad j = 1, \ldots, d \; j \neq i$$

$$p \in P_{i-} \quad \text{if } |y_i| > |y_j|, y_i < 0 \quad j = 1, \ldots, d \; j \neq i.$$

In other words, the data points p is most "deviant" or "atypical" in the ith component in the positive (above average) direction.

Therefore, there are a total of $2d$ possible pyramids that the d-dimensional data set can be divided into. For example, in Fig. 3.6, there are two dimensions X and Y, with four pyramids Y^+, X^+, Y^- and X^-. We call the attribute with the maximum deviation the **pyramid variable**.

Hyperpyramids

A finer *DS* partition is obtained by subdividing the pyramids into hyperpyramids. A hyperpyramid is specified by the attributes with the two largest deviations. If we extend the 2-D example above to a third dimension Z, we can define hyperpyramids of level 1. A point in the hyperpyramid $Y^+ X^+$ has the maximum deviation in the Y attribute and is above average, and the next highest deviation is in the X attribute where it is again above average. There are $2d(2(d-1))$ possible hyperpyramids of level 1 in d-dimensional space.

Level 1 hyperpyramids are a refinement of (level-0) pyramids, so level 1 hyperpyramids can be readily represented in a data cube by using a hierarchical dimension. Level 2 hyperpyramids are a further refinement of level 1 hyperpyramids, and can be represented by a three-level hierarchical dimension. In our experience, one level of hyperpyramids is quite sufficient for most analyses. However, the outer layers of heterogeneous data can sometimes benefit from a level-2 hyperpyramid. Here the benefit of using OLAP software is clear: the level of partitioning refinement can be easily manipulated.

3.4.3 EDM Summaries

Every layer-(hyper)pyramid combination represents a class in the *DS* partition. The data points in each *DS* class are summarized by statistics that are aggregable, that is, summaries of combined classes can be computed from the existing summaries without going back to the raw data. Some examples of aggregable summaries are counts (which we use in histograms), sums, sums of squares and approximate quantiles. Summaries are computed for all the attributes, not just the depth attributes. Attributes that are summarized but do not contribute to the depth computation are called **profiled** or **predicted** attributes, depending on their role. We will discuss predicted attributes later on in the chapter. The summaries need to be aggregable to facilitate rolling up the partition into a coarser partition (minimal partition computation) without re-visiting the raw data. The summaries are used for further analysis (set comparison, visualization, approximate nonlinear models). Note that we can keep a sample of data points from each class as an EDM summary. The sample can be used for the kinds of analyses (e.g., clustering) that cannot be performed with aggregates.

3.4.4 Annotated Bibliography

A center-outward approach to depth computation is discussed in [82] and [83]. See [72] for a discussion of the choice of center and its effect on the *DS* partition. Data pyramids were introduced in [7]. Finer partitions of data pyramids called hyperpyramids are discussed in [72]. A one-pass, efficient algorithm for computation of depth quantiles is presented in [85].

3.5 SET COMPARISON USING EDM SUMMARIES

We are often faced with the question of comparing two data sets. Are this week's network statistics different from last week's? If so, what is causing the difference? Is it very long, data-intensive sessions or relatively short bursts? There are many ways of comparing small data sets or univariate cases. We can compare means, quantiles, and so on. However, it is meaningless to compare the means (or any point estimate) of data sets that have millions of records.

The analysis is at too high a level and the interesting variations and details get swamped by the sheer volume of "typical" data. We discuss below set comparison based on a class-by-class testing technique.

3.5.1 Motivation

The comparison of two data sets can reveal a great deal of information. It can identify groups that are statistically different. For example, when it comes to buying automobiles, do women have different buying preferences from men? In general, we can use set comparison techniques for comparing customer behavior by region, by group, by month; comparing the output of automated data collecting devices to establish uniformity; Another important application lies in detecting inconsistencies between data sets. An automated fast screening of data sets for obvious data integrity issues could save valuable analysis time. Other applications include data process flow management and monitoring. An unexpected change in a data set can indicate a problem in the data collection process. We discuss the glitch detection aspect of set comparison techniques using a case study in Section 5.2.4 of Chapter 5. Set comparison can be also used to detect unexpected shifts in data over time. For example, suppose that the points in a data set represent a customer's activity measured through many attributes such as intensity of usage, duration, frequency of use and others. A comparison of the distribution of points in data sets collected at different points in time can indicate how the customer activity has changed between the observation periods. We will discuss how to tell genuine trends and shifts from spurious glitch related shifts.

However, detecting differences in the structure of large, high-dimensional data sets is a difficult problem. Furthermore, the classical multivariate statistics approach has centered heavily around the assumption of normality, primarily due to analytical convenience in terms of closed-form solutions. When the data sets are large, assumptions of homogeneity (e.g., i.i.d. observations) that underlie classical statistical inference might not be valid.

We have found DataSphere EDM summaries to be useful in performing nonparametric, piecewise statistical testing to identify sections where two (or more) data sets differ from each other. In principle, the following discussion is applicable to any partition, including axis-aligned linear and nonlinear partitions.

3.5.2 Comparison Strategy

In order to compare two or more data sets, we need to define a metric for similarity of two or more data sets. Similarity is tested in a multi step fashion (see Figure 3.7), using just the EDM summaries, which speeds up computation significantly. First, we test the distribution of points among the DS partition classes, within each subpopulation (combination of cohort values or equivalently every slice of the data cube). Second, we compare the multivariate means of the points that fall in each DS partition class. In addition, we use univariate

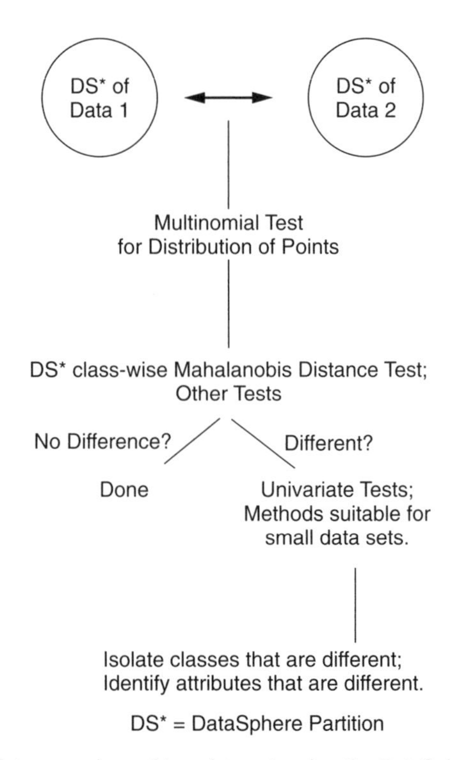

Figure 3.7: Set comparison of two data sets using the DataSphere partition.

tests for each variable individually to see which variable is driving the difference. In particular, we are interested in answering the following questions.

1. Which subpopulations have changed their behavior in Data 1 as compared to Data 2?
2. Of the subpopulations which changed their behavior, which sections show the most change?
3. Of the subpopulations which changed their behavior, which variables exhibit the most change?

3.5.3 Statistical Texts for Change

We use two complementary statistical tests that use only the EDM summaries. The first test is the *Multinomial test for proportions*, which compares the proportion of points that fall into each DS class. The second test is the *Mahalanobis D^2* test, which we use to establish the closeness of the multivariate means of each DS class within each subpopulation, for the two data sets. We use the distance between the multivariate means as a measure of similarity.

Note that it is sufficient but not necessary that the joint distribution of the two data sets be the same to pass these tests, and hence we use a strategy of multiple tests. In addition, one of the EDM summaries could be a sample of data points from the class. We can then use the EDM sample to implement nonparametric tests for shifts in distribution.

3.5.4 Application—Two Case Studies

In this section, we describe briefly two applications to illustrate the possible types of analyses.

We applied the DataSphere technique to data sets obtained from a corporate data warehouse. The first data set describes customer interactions with a service offered by the corporation. The data consists of over six million observations collected at nine evenly spaced intervals (approximately a month). Every data point consists of two cohort variables (VarA and VarB) related to types of service and tenure of a customer and six quantitative variables (Var0 through Var5) related to the usage of the service. The data set is divided into subpopulations based on the levels of the cohort attributes. Each subpopulation is subdivided into 10 layers using depth quantiles and each layer into 12 pyramids. A profile of EDM summaries is computed for each such layer-pyramid class. We expect that in aggregate, the customers in a subpopulation should have the same behavior from month to month. Any differences are worth investigating.

We analyzed the data set by applying a DataSphere to each successive pair of monthly data. Figure 3.8 shows the distribution of data points for a particular subpopulation among different layer-pyramid classes. The X-axis indicates increasing layers (layer 1 is the innermost layer, while layer 10 is the outermost) and negative layers indicate negative pyramids, and the Y-axis indicates different pyramids. The size of a circle is proportional to the number of points that lie in the indicated section. Note that the points are not distributed evenly among the layers because we plot the distribution of a pyramid and subpopulation, which constitutes a subset of the data points in a layer. Also note that creating a partition based on two criteria (distance layers and directional pyramids) enables us to *visualize* the aggregates of six-dimensional space in two dimensions.

We compared data sets from successive observation periods to determine if customer behavior changes between periods. In Figure 3.9 we show the results of the multinomial test. We plot a dot whenever the multinomial test indicates a significant difference. The purpose of this **bead plot** is to identify subpopulations that have a different distribution of data points among layer-pyramids, as compared to the previous month. Note that the beads are concentrated at lower end of the "age" (the tenure of the customer with respect to a particular service) variable. That is, in general, *new customers change their behavior, then settle into a more stable pattern*. Such information is useful for determining how best to provide service to the corporation's customers.

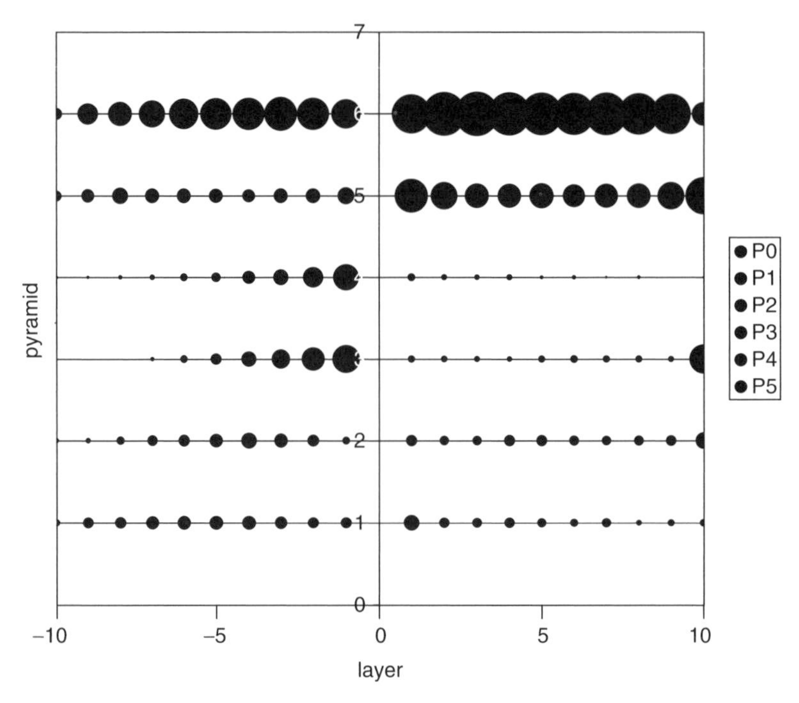

Figure 3.8: Distribution of points from the customer data set among sections. Dot size indicates the number of points in the section.

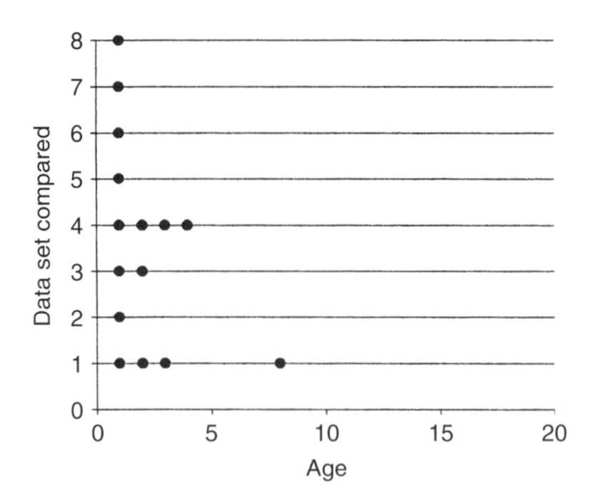

Figure 3.9: Multinomial test on the customer data for a given value of VarA. Dots indicate significant differences.

For the next stage of set comparison, we select two data sets where the subpopulations differ and apply the more detailed multivariate means test on a class-by-class basis. We plot the results in Figure 3.10, where the dots again identify layers that are statistically different. This comparison also shows that

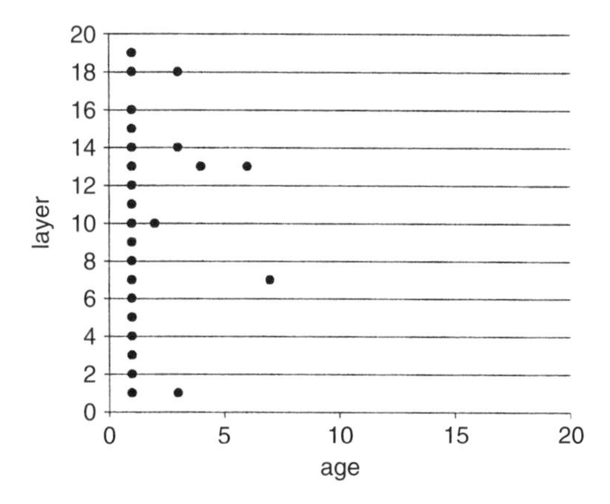

Figure 3.10: Multivariate means test on the customer data for a given value of VarA. Dots indicate significant differences.

recent customers change their behavior (from one month to the next) almost everywhere in the attribute space (e.g., they increase their usage, log on more often, etc.), while older customers are more stable.

Our second case study is a description of network traffic. The data set consists of 600,000 data points with two cohort variables (VarA and VarB) and three quantitative variables. Again, we give a high-level description to capture the nature of set comparison, without disrupting the narrative flow too much.

We collected two data sets from two different observation periods. We are interested in determining if network traffic is different at different observation periods. Our analysis indicated that there is no significant difference between the data sets collected at different time periods, *indicating that the characteristics of network use are stable over time.*

Next, we compared the network use for different subpopulations based on the cohort variable A. We found some significant differences. In Figure 3.11 we plot the distribution of points of a particular pyramid across the layers in the pyramid, for four subpopulations. The distributions of points for the subpopulations show the same basic shape, but some subtle differences. For a more refined analysis, we show the multivariate means test in Figure 3.12. Significant differences show up only for a particular pyramid P0- corresponding to low values of one of the numeric attributes (e.g., very short bursts of traffic). Hence we conclude that the subpopulations differ only in that section of the data that has exceptionally low values of the particular attribute. We have thus isolated subpopulations and the specific sections within the subpopulations that are different.

By partitioning the data sets, we can use well-developed nonparametric statistical tests to compare data sets. Data set comparison can yield interesting and useful information, indicating whether it is changing, and how. In Section 5.2.4 of Chapter 5, we will use this technique to detect glitches and learn how to distinguish genuine changes in data sets from data quality issues.

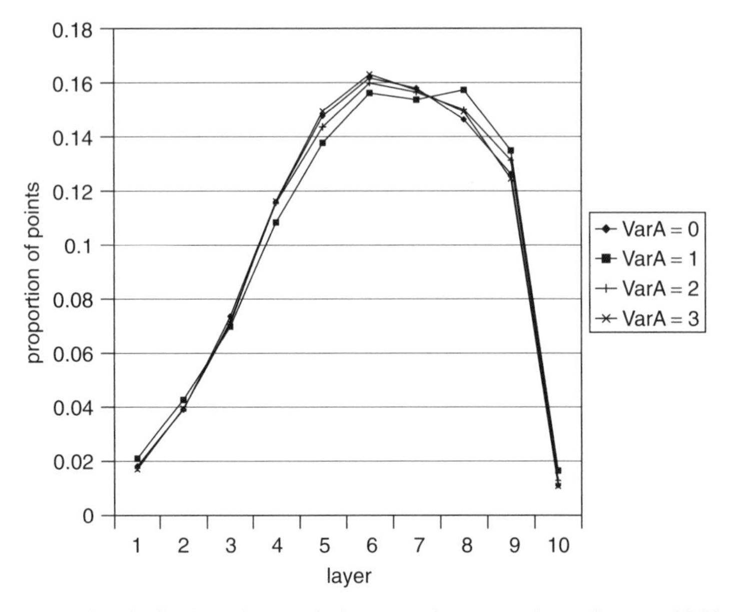

Figure 3.11: Distribution of network data set points across layers in pyramid P0-.

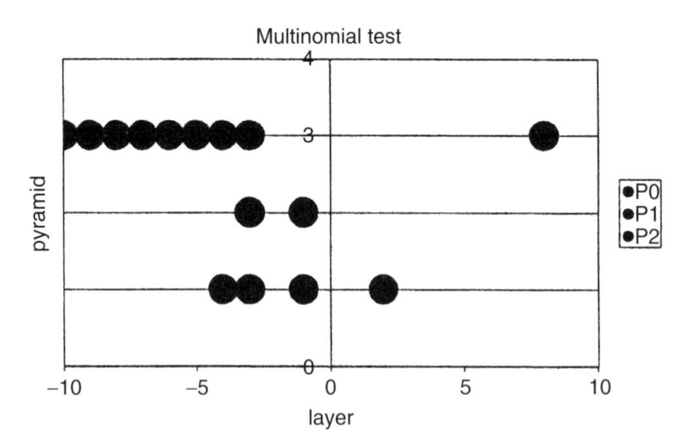

Figure 3.12: Multivariate means analysis of the network data, comparing data sets with VarA = 0 and VarA = 2. A circle indicates a significant difference.

3.5.5 Annotated Bibliography

Statistical tests like the Multinomial test for distribution of points in classes and the Mahalanobis test for distance between multivariate means are discussed in [105]. The book is also a good reference text for hypothesis testing, statistical estimation and linear statistical inference. A detailed description of set comparison using DataSpheres is in [71].

3.6 DISCOVERING COMPLEX STRUCTURE IN DATA WITH EDM SUMMARIES

In addition to set comparison, we are often interested in expressing relationships between attributes in a concise, compact fashion. For example, "every additional year spent studying increases the potential earnings of an individual by $20,000 annually." Or, "smokers are twice as likely to die of lung cancer than non-smokers at any given age." Such an exercise helps in prediction and is called **model fitting** or **model selection**. Model fitting is a way of compactly representing the structure in the data (data summarization, data reduction). There are many methods and disciplines, ranging from statistical models like log-linear models (logistic regression), proportional hazards to computational models like neural networks, clustering and hierarchical Bayesian networks. The common idea is to express the structure in the data set through a relatively small set of rules or specifications about the data and attribute relationships, that characterizes a major portion of the data with an acceptable degree of accuracy ("goodness-of-fit").

3.6.1 Exploratory Model Fitting in Interactive Response Time

Model fitting for large, heterogeneous, multidimensional data sets is difficult. As mentioned earlier, the data sets typically consist of a mixture of potentially large number of probability distributions so that most of the assumptions underlying existing techniques do not hold (e.g., "data are drawn from a distribution that belongs to the exponential family"). While these methods can be generalized to a mixture of distributions in principle, computing them involves too many parameters. That is, no single compact model with a reasonable number of parameters can accurately represent a major portion of the data set. Furthermore, representing complex nonlinear relationships is computationally expensive (e.g., clustering, classifiers) even for moderately large data sets.

An effective solution is to break up the data into smaller pieces and fit *fast, simple, approximate* models locally within each piece using EDM summaries. Such an approach is fast since it is based on summaries and enables us to *interactively explore* a large set of modeling options. The collection of approximate models represents the composite model for the entire data space. Such a composite model can be tailored to approximate the "true, arbitrarily complex model" to a desired degree of closeness by varying the granularity of the partition used to divide the data into pieces. However we should be careful not to overfit in which case we will be modeling the noise and not the structure in the data. The partitioning scheme use to generate the pieces of the data should result in a manageable number of classes, else the number of models to fit and store will be very large.

We discuss below two examples of piecewise fitting of approximate models. The first example is about approximating traditional statistical approaches where we are interested in answering a particular question like "how is a particular kind of response (buying brand A, switching telecom companies, amount of money spent on entertainment) affected by observable attributes (income level, education, number of telecom competitors in the area etc.)". Such questions are answered using statistical models that are broadly called **regression type** models.

The second example is more in the spirit of nonparametric approaches such as classification and clustering. We improve classification models by fitting simple one pass classifiers locally within each class of a partition of the covariate (feature) space, so that the attribute interactions are captured indirectly. Note that the common theme in the two approximate modeling techniques is the use of simple EDM summaries to approximate complex structure in the data.

3.6.2 Annotated Bibliography

A good overview of linear, log-linear and other types of regression models is presented [86]. A discussion of survival models such as the proportional hazards can be found in [24]. Piecewise model fitting and associated issues of smoothing and boundary conditions are discussed in [42] and [48].

3.7 PIECEWISE LINEAR REGRESSION

Consider the case where we wish to estimate the effect of the vector of *covariates* $\vec{x}$ on the *response variable* Y, represented by $g(Y) = f(\vec{x})$, where g and f are functions. Parametric methods assume that $f(\vec{x})$ has a convenient reduced representation such as

$$g(Y) = f\left(\vec{x}, \vec{\beta}\right)$$

where g and f are known functions and $\vec{\beta}$ is a vector of *parameters*. The problem is reduced to estimating just the parameters $\vec{\beta}$. Log-linear models are examples of parametric models. Computing the model parameters (e.g., maximum likelihood estimates) requires several iterations over all the data, using expensive algorithms such as Newton-Raphson or EM. If the assumptions are true, the fitted parametric models can be powerful, accurate and can be fitted using very little data. The problem becomes much harder when there is no prior knowledge to facilitate distributional assumptions or model selection. Ad hoc exploratory methods are an unsatisfactory option.

Therefore, using EDM summaries (which often include sufficient statistics) to construct fast nonparametric piecewise models is very attractive if we believe that such nonlinear relationships exist in our data. Furthermore, the

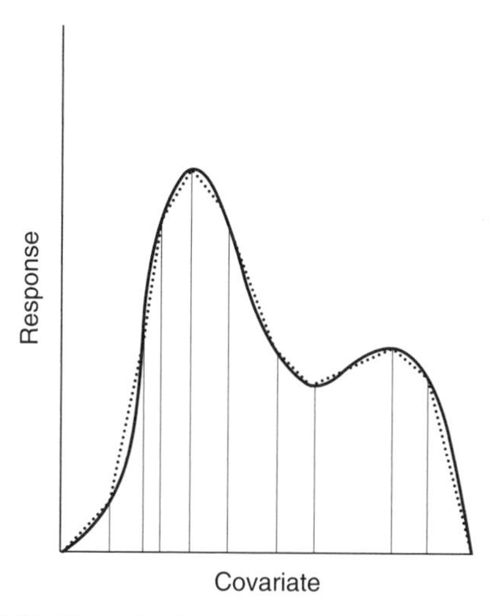

Figure 3.13: Piecewise linear regression with one covariate.

nonparametric models can often be leveraged to identify a suitable a set of parametric models that can be built using an EDM sample of the data.

A particular type of relationship that is easy to estimate from EDM summaries is **linear regression**, where the function $f(\vec{x}, \beta)$ is linear in the βs. That is,

$$g(Y) = \beta_0 + x_1\beta_1 + x_2\beta_2 + \ldots + x_d\beta_d + \varepsilon,$$

where β_s are the regression parameters, x_is are the attributes or covariates, and ε is a random error.

Linear regression simply means that the effect of certain attributes (*covariates*), on an attribute that we are trying to predict (*response variable*) is additive, as in:

$$Weight = 20 + 0.5Height + 0.75Age + \varepsilon. \quad (3.1)$$

In the above example, if we know the height and age of an individual, we can predict weight using the additive relationship. The quantities 20, 0.5 and 0.75, are called the **regression parameters or coefficients** and are *estimated* from the data using techniques like **least-squares**. We will not discuss regression methods in this book.

In Figure 3.13, there is just one covariate, so the piecewise model consists of splitting the single covariate into intervals and fitting a straight line within each. (Note that we need to impose constraints at the boundaries so that the two values from the neighboring straight pieces agree where we split the interval.) There is a vast amount of literature on smoothing, boundary conditions and choosing the intervals. When there are two covariates (example age and height), the covariate space is two-dimensional, so that our partition consists of rectangles rather than intervals. In higher dimensions, as we discussed earlier, the number of pieces in a rectilinear partition increases exponentially with a corresponding increase in the number of models to fit and to store. DataSpheres are a simple way to partition the covariate space into a manageable number of pieces to fit the models.

3.7.1 An Application

In this section, we give a brief description of the application of piecewise linear regression. We focus on conveying the potential use. The analysis is based on a real data set from a leading telecommunications company data warehouse. It has two objectives:

1. First, to use simple EDM summaries to construct regression parameters and use them to identify interesting attribute relationships for more rigorous testing ("every year adds 0.5 pounds to an individual's weight").
2. Second, to quantify the gains of using a piecewise regression model based on the DataSphere partition over fitting one single regression equation to all the data. We would expect the piecewise model to do better since it is reflects the local variations in the data better.

The data set consists of six quantitative variables measured on 947,711 individuals. The variables measure different aspects of an online service offered to consumers. In addition, two cohort variables corresponding to a subscription date and type of service are associated with each customer.

We are interested in expressing the relationship between the predicted variable, which measures the duration spent using the service, in terms of five covariates, which are measurements of different types of activity generated while using this service. We used the five covariates as *depth variables* (from our DS terminology), to create a DataSphere for every combination of the levels of the categorical variables (i.e., *cohort variables*). The EDM summaries from every layer-pyramid combination are then used to compute the linear regression coefficients for relating the *predicted* variable to the five covariates (X_1 through X_5) using the *least squares method*.

3.7.2 Regression Coefficients

Briefly, the picture in Figure 3.14 plots the regression parameters (βs) of three of the covariates for each class of the DS partition. The plot clearly shows that there is enough variation in the coefficients to justify a piecewise fit using the DataSphere partitioning scheme. Similar variations can be seen in coefficients corresponding to other covariates as well.

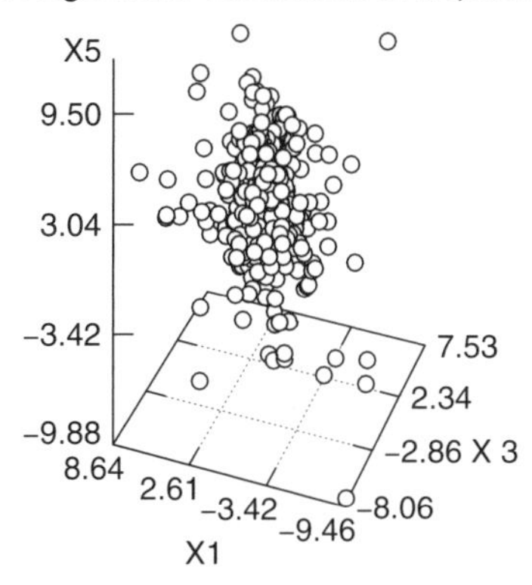

Figure 3.14: Variation in regression coefficients of $X1$, $X3$, $X5$.

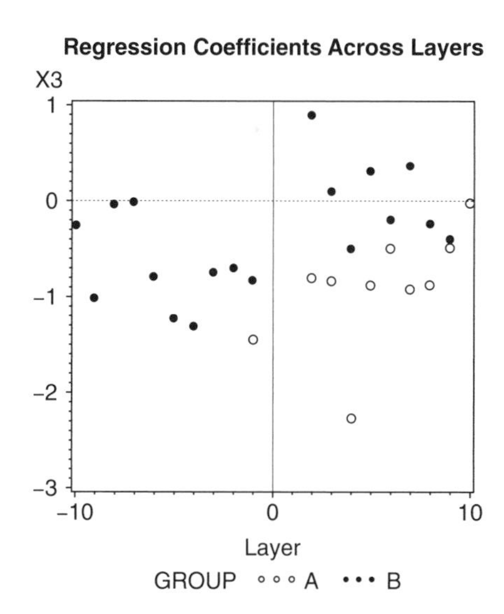

Figure 3.15: Regression coefficients of $X3$ across layers.

Next, consider Figure 3.15, which shows the values of the regression parameters (β) for one of the covariates for two subpopulations (black dots, white dots), corresponding to two different types of online services. The regression parameter is plotted against the layers, where a negative layer corresponds to a layer in a negative pyramid of a particular attribute. It is clear that the coefficients of the two subpopulations A and B occupy different regions. The two groups can be distinguished or separated based on their regression coefficients. The statistical significance of the difference in the coefficients of the two groups needs further testing. We have used the approximate model based on

EDM summaries to rapidly identify subpopulations that are potentially different. The suspect subpopulations, being significantly smaller than the entire dataset, can be investigated further using more computationally intensive methods if necessary.

3.7.3 Improvement in Fit

We computed the *R-square* for each of the classes in the DS partition. ***R*-square** is a measure of goodness-of-fit and represents the amount of random variation in the predicted variable associated with the variation in the covariates. We give an informal definition below, where Y is the response variable:

$$R^2 = 1 - \frac{\sum (Y - \hat{Y})^2}{\sum Y^2} \tag{3.2}$$

for a given section. Here, $\hat{Y}$ represents the predicted value of the response variable Y, obtained from the regression equation for each class. We will use the difference in R-square as an informal measure of difference in fit, in tune with the exploratory nature of the DataSphere analysis. We restricted ourselves to sections in the partition that have a minimum of 10 observations. We noticed that the sections with the maximum improvement in R-square were those that have a small number of observations and were often the outer layers. This indicates that the coefficients are considerably different from the overall fit. That is, we have isolated subgroups that are different from the overall population and need separate models.

In this section, we have shown that the EDM summaries based on a suitable partition provide a fast way of fitting piecewise regression models for prediction. We have not included discussion of smoothing the piecewise models and boundary conditions since this is a more specializes branch of analysis that is outside the scope of this book.

Note that the approach is applicable to other nonlinear models like *proportional hazards* and others where we can find simple nonparametric models that use EDM summaries and provide the same functionality as parametric models, are much faster and more widely applicable. We have successfully used the nonparametric Kaplan-Meier estimate of the survival function within every class of the DataSphere partition to approximate complex nonlinear models like the proportional hazards model.

3.7.4 Annotated Bibliography

A general reference for linear models is [86]. The book also contains a discussion of the Newton Raphson method. The EM algorithm is discussed in [35]. An introductory, intuitive explanation of regression, random variation, estimation through least squares and the R-square measure of fit can be found in [47]. A linear algebraic treatment of linear regression with discussion about error assumptions is found in [105].

A discussion of smoothing, boundary conditions and other matters related to piecewise models can be found in [42] and [48].

For an introduction to survival analysis and the proportional hazards model, see [24]. An example of approximating a nonlinear model like the proportional hazards with a collection of piecewise estimates of the survival function is presented in [31].

3.8 ONE-PASS CLASSIFICATION

An important focus in EDM is to *reach an approximate solution relatively fast*, rather than find an extremely accurate solution that takes hours if not days to complete. While this is contrary to the classical statistics approach, it is a necessity dictated by the volumes of data (e.g., streaming data) that get seriously backlogged if not dealt with rapidly. In this paper, we propose scaling a popular data mining technique, *classification*, to large multivariate data sets by fitting simple one-pass classifiers within each class of a scalable partition of the attribute space. In addition to speed and flexibility, the technique has the advantage that the assumptions of independence of attributes that underlie the simple models (Naive Bayes) hold locally within each class of the partition. The variation in the classifiers across the classes captures the interaction of the attributes.

3.8.1 Quantile-Based Prediction with Piecewise Models

Consider the problem of binary prediction. Given a binary response variable Y, we are interested in predicting its value as a function of the covariate vector $X = (X_1, \ldots, X_d)$.

A simple model can be built based on the conditional distribution of X_i given Y. An example is Naive Bayes classifier, which uses the Bayes rule for prediction. That is,

$$P(Y|X) = \prod_{i=1}^{d} \frac{P(X_i|Y)P(Y)}{P(X_i)},$$

where the covariates X_i are assumed to be mutually independent. Continuous covariates require further assumptions about the form of $P(X_i|Y)$. We take a more general approach by "discretizing" the X_i's using quantiles.

Furthermore, we make the above simple model more effective by applying it within each class of a partition of the covariate space. Axis-aligned, rectilinear grids based on **equi-depth marginal histograms** (each bin has same mass) of the covariates are not suitable because number of classes in the par-

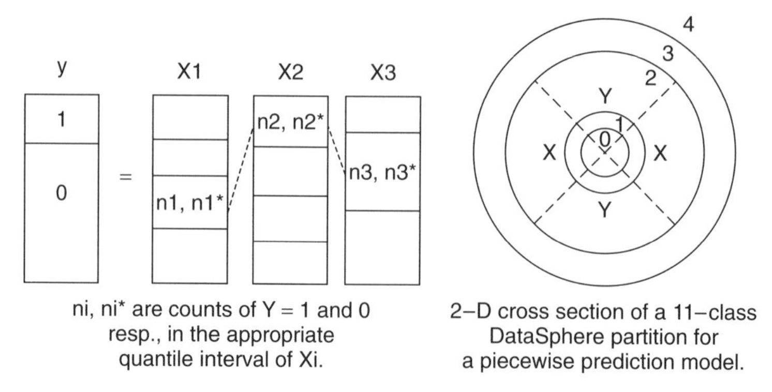

Figure 3.16: *Left*: Univariate prediction, independence assumption; *Right*: Multivariate, piecewise prediction.

tition increases exponentially with the number of dimensions. Within each class of the DS partition of the covariate space, we apply the simple binary prediction model with quantile based discretization of the covariates. The interactions between the covariates are reflected in the variations in the predictive models across the classes of the partition.

The simplest method of estimating $P(X_i|Y)$ is linear interpolation within any bin of the histogram, assuming that all values are equally likely within the bin. More sophisticated estimates can be computed using **kernel splines** or by making other assumptions about distribution of points within a bin. For discrete variables, we simply take the proportion of all sample points that fall in the bin.

In general, the overall likelihood ratio is computed via

$$l(X) = \frac{l(Y=1 \mid X)}{l(Y=0 \mid X)} = \prod_{i=1}^{d} \frac{P(X_i \mid Y=1)}{P(X_i \mid Y=0)},$$

after simplification. (See Figure 3.16, left.) We ensure that the estimates are robust by choosing quantile based intervals that contain a guaranteed number of points.

The process of prediction simply consists of choosing a cutoff γ such that we predict $Y = 1$ if $l(X) > \gamma$ and 0 otherwise. Note that γ is a parameter of the model, whose choice can be determined by trading off the different types of errors, such as Type I and Type II. γ will vary across the partition.

3.8.2 Simulation Study

We created a data set where the response variable Y was 1 if:

$$2x - y^2 + 0.5z^2 + 3xy + 5yz + xyz >= 1$$

and 0 otherwise, where x,y,z are standard Normal variates. The data set consisted of 200,000 observations with 124,692 responses being zero, so that approximately 37.66% of the responses were 1.

An application of the likelihood ratio based prediction using marginal distribution of the covariates, to the entire data set resulted in a 13.09% misclassification error. We then created a DS partition of the covariate space x, y, z into five concentric distance layers and fitted a binary prediction model within each layer.

5-Layer Partition			Layer-Pyramid Partition			
Layer	N	% Error	Layer	Pyramid	N	% Error
0	39,984	16.5	0	0	10,795	5.45
1	39,990	18.6	0	1	13,878	7.4
2	39,969	15.7	0	2	15,311	15.51
3	40,083	10.4	1	0	12,902	7.2
4	39,974	8.3	1	1	12,743	14.54
			1	2	14,345	13.76
			2	0	12,812	8.5
			2	1	12,655	11.76
			2	2	14,502	8.31

The central layer around the center is referred to as the 0 layer. Since layers 3 and 4 seem to have a lower error rate, we ignored them and further partitioned the layers 0,1,2 into 3 pyramids, by dropping the distinction between the positive and negative pyramids of any attribute. The overall error rate is now 10%. That is, even with a very crude partition of 11 large classes ($N \geq 10K$), we have a significant reduction in the error rate. Note that we can selectively refine the partition to focus on regions where the fit is poor. However, we have to be cautious not to overfit.

We have shown that a fast, effective way of modelling a complex multivariate data set is through quantile based, scalable partition such as the DS partition which serves as a grid for fitting piecewise models. Furthermore, we have shown that a simple quantile based binary prediction model based on assumption of independence of covariates can be extended to capture covariate interactions by applying it within each class of the partition.

Further refinements can be made by exploring: (1) The effect of highly correlated variables—correlated attributes will reinforce each other and dominate the prediction. It is enough to keep a single representative attribute. (2) Other options for variable selection, and (3) Different types of probability estimation to improve the model while preserving simplicity, computational speed and ease of interpretation.

3.8.3 Annotated Bibliography

See [52] for a discussion of statistical modelling with quantile functions. See [36] for a discussion of Naive Bayes classifier and the robustness to the failure of the independence assumption.

3.9 CONCLUSION

In this chapter we have introduced data partitions such as data cubes, DataSpheres and others that are induced. We have used the partitions to create fine grained summaries of the data which we call EDM summaries. In addition to revealing simple characteristics of the data, the EDM summaries can be used to make comparisons of groups and to detect trends over time. Furthermore, EDM summaries are very effective in rapidly detecting structure in the data such as inter-relationships between variables. The results from such exploratory model fitting can be used to interactively select more accurate, sophisticated models which can be computed using a very small subset of the data. Alternately, we can construct piecewise models with simple components using EDM summaries to approximate complex nonlinear models. EDM summaries can also be used to detect anomalies in data sets.

CHAPTER 4

Data Quality

4.1 INTRODUCTION

Data quality is a complex and essentially unstructured concept. Many disciplines have taken one-shot approaches addressing simpler abstract versions of the real problem. A major challenge in devising general solutions is that solving data quality problems requires highly domain-specific and context-dependent information, involving interaction with domain experts. Only experts can specify the rules and data flows (dynamic constraints) that are correct. Developing such a set of rules is a critical step in data checking and validation. In addition, the set of specifications that define appropriate data behavior are skewed, in the sense that, while a handful of rules can specify say 50% of the data, every additional rule specifies smaller and smaller portions of the data. To cover 90%–95% of the data, we might need hundreds of rules, coming down to a case-by-case basis in the last 10% or so of the data. We will explain this in detail in Sections 4.4.6 and 5.5. The ultimate goal of DQ methods as well as metrics is the improved *usability* and *reliability* of the data.

Data quality monitoring is an incessant and continuous activity starting right from the data gathering stage to the ultimate choice of analysis and interpretation of the results. We need to update the static conventional definitions and metrics of data quality to reflect the continuous and flexible nature of the DQ process and metrics needed to effectively measure and monitor data quality. Consider the following two remarks:

1. "Current data quality problems cost U.S. businesses more than 600 billion dollars a year."
2. "Between 30% to 80% of the data analysis task is spent on cleaning and understanding the data."

Exploratory Data Mining and Data Cleaning, by Tamraparni Dasu and Theodore Johnson
ISBN: 0-471-26851-8

The first remark is based on a study conducted by The Data Warehouse Institute, commissioned by DataFlux [122]. The second remark is survey based, from conversations with practitioners of data mining. Factual or anecdotal, the importance of data quality is becoming increasingly clear, as evidenced by the surge in software, tools, consulting companies, and seminars addressing data quality issues. In addition to the cost in terms of time and resources, there are hidden costs to poor corporate data. Bad data can lead to inaccurate bills, resulting in loss of revenues (underbilling, losing customers through overbilling), loss of credibility ("If they cannot bill correctly, how can they provide 24/7 network reliability?") and an immense overhead in customer care. In addition, inaccurate databases can hamper the provisioning of new services. For example, customers might want to manage their own accounts in terms of changing the products and services they buy (e-management and e-provisioning). However, in order to offer this convenience, the databases need to reflect accurate and current views of the customer's account.

In this chapter we discuss data quality as a continuous concept from the data gathering stage to the ultimate analysis stage. We call this the *data quality continuum*. Each stage has its unique source of problems that need specific solutions. Therefore, we discuss defining data quality in the context of the data quality continuum. Such an approach requires updating conventional data quality definitions to suit the vastly changed data universe today.

An important aspect of the data quality continuum approach is the interaction with the data. Data quality is an iterative process where data mining (EDM) techniques can help unearth data glitches. As discussed earlier, the two are closely related. The top results of a data mining exercise are certain to include at least one data glitch or data misinterpretation. Data misinterpretation occurs when the metadata are insufficient to fully interpret the data. For example, "the attribute Income represents the after-tax income, except in the state of New Jersey, where it represents the pre-tax income." If this rule were not known, New Jersey residents would show up as extreme outliers in a data mining exercise. An EDM phase is often an effective way to unearth inconsistencies that point to potential hidden data rules. *Data browsing* and *data profiling* are effective EDM methods that help "discover" rules and specifications that were not officially documented, in addition to starkly highlighting the flaws in the data.

Consider the following example, where we don't know much about the data and need to discover a unique identifier to match the data set D1 with other data sets.

- Data Set: *D*1
- Number of data points: 1000
- Attribute: A
  Unique values: 4
  Missing Values: 3

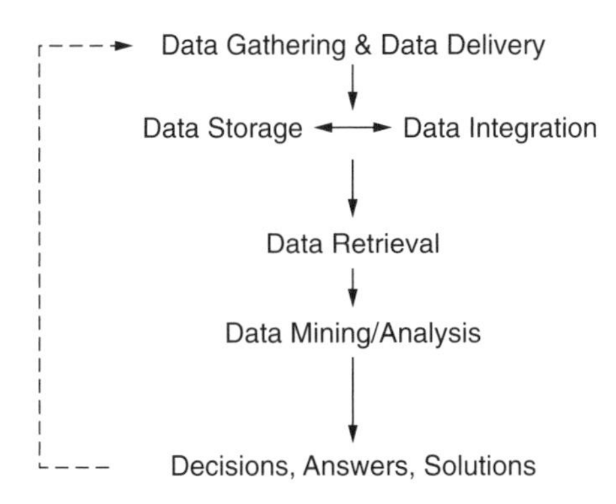

Figure 4.1: Data quality watch points.

- Attribute: B
  Unique values: 980
  Missing Values: 0
  Frequency Tables:
  "Null" 6
  "0" 5
  "Blank"5
  "000-000-0000" 5
  "N/A" 4
  "123-456-0000" 1
  "123-456-0001" 1
  . . .
- Attribute: C
  Unique Values: 10
  Missing values: 950

Attribute A is well populated but has only 4 unique values. It is unlikely to be a unique identifier (match key). Attribute B is well populated and has a large percentage (98%) of unique values. However it has a small amount of corruption in it in the form of nulls and junk strings like "000-000-0000". Once the junk is removed, attribute B is a good candidate for a match key. Attribute C is fairly useless since it is not populated most of the time. It is clear from this simple example that data browsing and profiling can provide insights and help identify obvious data quality issues. In the next chapter we will discuss tools and algorithms that help with data browsing aspect of EDM.

In this chapter, we will give a brief overview of four complementary approaches to data quality derived from the areas of:

1. Process;
2. Statistics;
3. Database; and
4. Metadata and domain expertise.

The statistical and database approaches are frequently *diagnostic* in the sense that they happen after data collection. We discuss these two approaches in detail in Chapter 5. The process and metadata/domain expertise aspects are integral to the design of data gathering, data flows, data specification, data storage, data retrieval, data analysis and data monitoring. We consider these to be *pre-emptive* approaches to data quality and discuss them in this chapter. Wherever possible, we integrate case studies into the narrative.

Previous texts have discussed static and context-free definitions of the meaning of *data quality*. In this book, we explore a definition of data quality which relies on the nature of the data and its intended use. Finally, we propose a method for developing **data quality metrics**, based on defining a set of *constraints* that the data should satisfy. These constraints are of two types: **static constraints**, which match the data to its schema or other descriptions, and **dynamic constraints**, which match data flows to business rules.

4.2 THE MEANING OF DATA QUALITY

The meaning of data quality is often closely tied to the context and application. However, there are certain conditions and constraints that are common to most data sets. In this section, we will discuss both types of data quality (DQ) considerations.

4.2.1 An Example

Before we discuss the meaning of data quality, consider the following hypothetical example:

$$T.Das|97336o8327|24.95|Y|-|0.0|1000$$

$$TedJ.|973-360-8779|2000|N|M|\mathrm{NY}|1000$$

The data represent two records with pipe ("|") delimited fields.

- **Can we interpret the data?**: Without a data dictionary the above data are unusable. It does not matter even if the data are accurate, timely, consistent and complete. While we can make some educated guesses about the meaning of the fields, the lack of sufficient data descriptions are a leading cause of poor data quality. Next, suppose we get hold of a data dictionary. It tells us that the fields are name, phone number, revenue, indicator, gender, state and usage.

- **Data quality problems**: At the very first glance, we can identify some data problems. The name field has two different representations "First initial Last name" and "First name Last initial." There is no *standard way* of representing the name attribute. If there are no standards across and within a database, joining tables to create integrated views becomes a huge problem since there are so many exceptions that need to be considered. A similar problem exists with the phone number which is hyphenated in the second record but not in the first. In addition, note the typo of "0" as "o". The possible reason could be **manual entry** of data. Also note that the Gender attribute is *missing* in the first record. Another *inconsistent entry* is the State attribute in the first record. It should be a string but what does "0.0" mean? We could find the above obvious problems using the data description given. However, there are other problems of data parsing and data interpretation that are not evident. Such problems are revealed as a consequence of verifying the static constraints on the field formats.
- **Metadata and Domain Expertise**: There are hidden problems that require knowledge over and above a mere description of the data. For example, it seems that Ted J. generates revenues several magnitudes larger than T. Das. A data mining algorithm would pick Ted J. as an outlier. However, it is possible that T.Das' revenue was reported in dollars (convention in New Jersey state) while Ted J.'s revenues were reported in cents (convention in New York). Similarly, we would put both the users in the same usage category since the usage is 1000 for both. However, we were not given an important piece of information. If the indicator is "Y" then, the usage has been *censored*. That is, we know that the usage is at least 1000, we do not know if it is 1001 or 100000. We will discuss censoring in greater detail further in the chapter. From this example it is clear that there are rules and conventions related to process implementation that need to be used to interpret that data. Such rules are often obtained from *domain experts* and should be a part of *metadata*, data about the data. We will define these concepts in detail in Section 5.4. Major failures in data quality (data misinterpretation) often arise from insufficient metadata and domain expertise.

4.2.2 Data Glitches

A **data glitch** is any change introduced in the data by causes external to the process that generates the data and is different from the usual random noise present in most data sets. Noise is caused by uncontrollable measurement errors such as imprecise instruments, subtle variations in measurement conditions (normal wear and tear of hardware, software degeneration, climatic conditions) and human factors. Data glitches on the other hand are systematic changes caused by mega phenomena such as unreported or dropped data, unintended duplicate records, switched fields and so on. Some inconsistencies are obvious and easy to detect while others are subtle, and are noticed only after they have been compounded several times resulting in significant deviations from the true values, necessitating expensive backtracking. Localized errors are swamped in aggregates, and therefore go undetected for quite some time. We present below an anecdotal discussion of a subset of commonly encountered data glitches.

Unreported Changes in Layout

When data processing centers make changes, downstream users are not aware of these changes for a short interval of time. However, they continue to receive and use the data feeds during this interval. Some of these changes are obvious, such as a change in layout where the position of a 13 character string variable is switched with a float. But sometimes the changes are subtle (e.g., a swap of two numeric fields) affecting only a few variables that resemble each other in their univariate behavior but differ in their interaction with other important variables. These glitches can be hard to detect because univariate tests and aggregates will not detect such changes. An example of such a condition could be the switching of the fields that measure the customer usage of a service with two competing providers. While the overall patterns might be similar, each provider might be used for a different purpose, affecting its interaction with other variables such as time of day, application and others.

Unreported Changes in Measurement/Scale/Format

In some situations, a field or variable is sent to the users without being processed completely. This could be due to a program exiting without completing, yet generating no error message. The processing could be such that it affects only a small proportion of the records, but has serious consequences. An example is the application of volume discounts to generate customer bills. Some discounts are so structured that only a fraction of percentage of top users qualify. However, failure to process the discounts can create serious customer satisfaction problems alienating valuable customers. Again, tests based on aggregates often fail to detect such errors.

Temporary Reversion to Defaults

A third kind of frequently encountered glitch is caused by the defaulting of measuring devices to pre-set limits. For example, the reported length of any telephone call exceeding 100 minutes could be defaulted to 100, due to some temporary condition in the switches. Aggregates do not reveal such glitches unless the error condition persists for a prolonged period of time. However, the existence of such a condition for even short periods of time could result in lost revenues. Therefore it is important to detect such errors as close to real time as possible.

Missing and Default Values

Missing values are very prevalent in data sets. There are many different ways of dealing with them such as dropping them from analysis or substituting typical values for them. The approach depends on the amount of data missing as well as the nature of the application. An extra complication occurs when the missing values are defaulted to a valid value of the variable itself, usually an infrequent one. An example is representing missing values by zero, even though zero might be a valid but unlikely value of the variable. The implications can be serious if there is a sudden increase in the valid "zero" values, which will be masked by the missing "zero" values. While it is obvious that setting such defaults is incorrect, decisions for data collection and measurement processes are not necessarily made with a view to future analyses. Sometimes the limitations of the systems that process the data force such ambiguous defaults.

Gaps in Time Series Records

Discontinuities in historical or transactional records can be detected easily once the need for detection has been established. For example, in a system that updates the status of a data point, it is simple to verify that the update applied to the old status results in the new status. Consider the following sequence of updates:

1. current status = 3 cellular phones, 4 phone lines
2. update: drop = 2 cellular phones, add = 1 phone line
3. new status = 0 cellular phones, 9 phone lines.

Clearly, some intermediate updates are missing. However, the problem becomes serious when there are many such missing records and a large portion of the data set is quickly disqualified.

4.2.3 Conventional Definition of DQ

Traditionally, data quality has been defined and measured using stringent constraints that need to be satisfied. For example, the concepts of *accuracy, completeness, timeliness* and *consistency* figure frequently. We discuss these concepts briefly. **Accuracy** pertains to the closeness of the value in our database to the true value. Accuracy is difficult to measure since very often we do not know the true value. However some instances like "0.0" or "NX" for the attribute State are obviously inaccurate. **Completeness** measures how completely the target domain is represented in our database. For example, if we are trying to represent the Himalayan organisms, to what extent have we covered the population? Many data quality issues arise out of a lack of completeness. What if the census mistakenly leaves out the Asian or Hispanic populations in the United States? Poor performance on this metric can seriously bias the analysis performed using the data and the decisions thus made, even if the data are accurate. Completeness also is difficult to measure. **Timeliness** refers to the currency of the data. That is, the most recent time when it was updated. This DQ dimension is critical for certain attributes like weight that change over time, while it is irrelevant for attributes like specie type and gender that are fixed and do not change with time. A special consideration with respect to the time aspect is synchronization—ensuring that time windows match up (e.g., we can't roll up weekly summary to compare to monthly summaries without a significant loss of accuracy). **Consistency** ensures that the there are no conflicts within or between data sets. For example, we might find usage for a service that a customer has not subscribed to, or the same person might have different genders in different data sets. Many databases provide for automatic checking of inconsistency by defining *constraints* that the data values need to satisfy. Despite such provisions, inconsistencies abound in most real-life data sets.

There are other dimensions of data quality such as content, level of detail, appropriateness, interpretability and portability pertaining to conceptual views and representations of the data. We will discuss these issues along with additional concepts in the context of the *data quality continuum*.

4.2.4 Times Have Changed

The DQ metrics discussed in the above section impose rigid and somewhat static requirements on the data. However, we need more updated and flexible criteria for judging the quality of contemporary data sets. The nature of collection as well as the size, variety and content of data have changed dramatically in the last decade. In fact, the definition of data itself has changed dramatically to include any kind of information that is analyzed systematically. So have the expectations of data changed. In addition, we need to incorporate the domain-specific nature of DQ into the metrics by making them flexible and easy to customize. Some reasons for revisiting the definition of DQ metrics are:

- **New data paradigms**: Traditionally, data were collected from well-designed experiments tailored to answer specific questions. The measurements were often meticulous and repeated to adjust for random variability. Statisticians are accustomed to such data. However, things have changed dramatically. We now have automatic data collection systems that spew out monster torrents of data (we discussed summarizing such data in previous chapters). These data sets are often "found data"—being a byproduct of another process (e.g., web server logs), or extracted from a non-authoritative source (e.g., web-scraped tables). Therefore, such data streams are plagued by data quality problems. We cannot exercise any control over the design or collection of such data. In fact, the data collection mechanisms and conventions are frequently opaque so that *our understanding of the data is flawed*, introducing data quality concerns. In addition to the volume, lack of control and transparency, the type of data has taken on new forms as well. We now have images, web pages, web server logs, audio files and time series that arrive out of chronological order. A frequent source of DQ problems is the *inappropriate use of known data representations* to incorporate new data types.

- **Federated Data**: As corporations seek to use their data effectively, they try to make their data "information rich" by bringing together many different data sources. **Enterprise data**, where many disparate data are integrated, often by force, are highly valued. However, data integration creates its own DQ worries. Data sets often do not have common **match keys** to bring them together. For example, if data set $D1$ contains species type and weight and data set $D2$ contains species type and volume, how do we get a complete picture of each of the Himalayan organisms? We need an additional attribute in each of the data sets, say *Id*, that *uniquely* identifies each organism and is *the same in the two data sets*. We can then **join** the two data sets using the *Id*, constructing a combined data set $D3$ that has the complete information

  Id, specietype, weight, volume

  for each of the uniquely identifiable Himalayan organisms. Frequently, within the same corporation as well as when companies acquire each other, a clean common match key (or **join key**) is missing. We are forced to use approximate joins using imperfect match keys like name and address resulting in **approximate joins**. For example,

  Ted Johnson, 3 apples, 09-01-2001

  and

  Theodore Johnson, 2 CDs, 09-02-2001

  pertain to the same individual but cannot be integrated because the first name is spelled differently. Approximate matching algorithms (we discuss these in detail in Section 5.3.4) can give spurious matches with

  Ed Johnson, Eddy Johnson, Todd Johnson

  potentially different individuals. Such situations require the use of strong DQ checks, validation and controls.

- **Disconnect between data collection and analysis**: In the past, data were collected specifically with a particular objective in mind. The attributes needed were predetermined and measured. In fact, even the analysis was determined ahead of time and the data were measured accordingly. The area of **design of experiments** in statistics is dedicated to this topic. However the data sets that are "mined" and analyzed today are chaotic, entered by harried sales personnel, scraped from convenient but buggy sources, and/or generated as a byproduct of a process. There is no consistency or a standard representation for the same data attribute, as we saw in the example where revenues for T. Das were reported in dollars and revenues for Ted J. were reported in cents. Furthermore, the meaning and interpretation of the data are inadequately documented so that the data become unusable by the time the analyst gets it.
- **Great Expectations**: As the data sets have become more massive and varied, the expectations from the data have changed. No longer do we use data for merely creating sedate summaries and reports. We want to mine data, analyze it and use it as an oracle to prophecy trends, opportunities and risks. At one point, the vision was to have a "black box" that would ingest raw data and output actionable intelligence and rare insights. However, it is now accepted that data mining is an interactive process with several iterations incorporating feedback from experts, domain knowledge and insights from earlier data mining phases.

Given the dynamic nature of the data domains and technology environments, *the meaning of DQ and associated metrics should be revisited frequently* to incorporate the changing needs.

4.2.5 Annotated Bibliography

See [106] for a detailed discussion of the concepts of consistency, accuracy, uniqueness and timeliness and other process related issues. These constitute the static constraints. The book also contains a discussion of other dimensions of data quality. Dasu and Johnson [30] provide an extensive discussion of data glitches. See [81] for material on the treatment of missing values in Statistics. Fisher [46] is a good reference for the statistical design of experiments.

4.3 UPDATING DQ METRICS: DATA QUALITY CONTINUUM

In this section, we emphasize the continuous and pervasive nature of data quality measurement by discussing it in the context of the entire data flow process using the **data quality continuum**. The major stages in the data quality continuum are depicted in Figure 4.1. As mentioned earlier, the DQ problems that arise at each of these stages are different, and need different metrics as well as solutions.

4.3.1 Data Gathering

In increasingly rare cases, the data gathering stage is preceded by a **planning stage**, where the amount and type of data gathered are planned and provided for. More often than not, however, it is the constraints on measuring devices ("router can be polled only every one hour") and business exigencies ("the data base to support Product A has to be up and running by tomorrow") that determine the kind of data that is gathered, without the pre-planning stage.

Problems

As we saw in the example in the introduction, a major source of error at the data gathering stage is the **manual entry** of data, which can result in *mis-typed* data such as age is 90 instead of 19. Manual entry can also lead to incomplete and missing data as often happens when the data source's priority is to 'sell, sell, sell' rather than accurate data entry. It can also lead to *non-standard* entry like "Ted J." and "T. Das". Another frequently encountered problem is **duplicate entry** of data. The database community has several algorithms for merging and purging duplicates as we will see in Section 5.3.4.

The lack of a *pre-planning stage* often introduces intractable DQ errors such as a mismatch between the scale of the problem and the scale of the hardware/software used. It can lead to short sighted or uninformed decisions that constrain the functionality of the data. For instance, it might be decided to use daily aggregates rather than hourly aggregates of network traffic due to disk or memory constraints. However, this might make the data unusable for fine-grained analysis to study peak traffic patterns. Or the measuring device might measure only a maximum of 1000 *kilograms*, "censoring" all values above that limit. The database design might try save a few bytes through a special data encoding, which makes hidden assumptions about the data to be stored (e.g., the infamous "Y2K" problem). Finally, there is always the measurement error caused by inherent random variability that can be addressed using statistical techniques. We have recounted a small fraction of the potential errors that can be made during the data gathering stage.

Potential Solutions

There are two major approaches to handling DQ errors during data gathering—pre-emptive and diagnostic. A **pre-emptive** approach is based on process and architecture, to ensure the following:

- Design and manage the processes to automate the work flows as much as possible and avoid manual intervention and inputs.
- Enter data once and enter it right—interfaces should be designed to permit data entry in a standard form and to prevent duplicate entries.
- Emphasize data sharing and data maintenance. Any changes in the metadata should be shared with downstream users immediately. In fact, data and metadata should be inseparable.

- Assign responsibility to **data stewards**, who are subject matter experts, who know and understand the data and how they are used. They are charged with maintaining the quality of the data. U.S. government agencies such as the EPA use data stewards extensively.
- Conduct frequent end-to-end audits, starting at the data gathering stage, to catch data corruption as soon as possible. Such audits need to be continuous and could be hard to implement.

Frequently, there is no way to control or manage the processes. The data are what they are and the analysts and data miners have to deal with the data the way they are. Statisticians and the database community have developed **diagnostic** measures to hunt for data glitches in data sets and for cleaning them up. These include methods for merging and purging duplicates, mapping nonstandard formats to "tokens" that can be used across tables and data sets, set comparison techniques to identify sections of data that are suspicious and others. We discuss these techniques in detail in the next chapter.

4.3.2 Data Delivery

The process by which the data are sent from the place of origin to a permanent storage place constitutes **data delivery**. It is a complex process involving multiple files, multiple feeds that have cross dependencies and time synchronization issues. It takes rigorous design and monitoring to ensure data delivery without loss and distortion of the data.

DQ Problems During Data Delivery

There are two major categories of problems that arise during data delivery. The first has to do with *data mutilation* and the second concerns *data loss*. It is hard to say which is worse!

- **Data mutilation**: Often, the data are pre-processed before being sent to downstream users. Reasons for doing this include space constraints, memory constraints, time limitations, and sometimes a desire to hide information. Data are aggregated (e.g., by time, by demographic group) so that fine-level granularity is lost. This can be a problem if we want to analyze patterns in short intervals of time or if we want to analyze individual customer behavior.

  Another instance of data mangling is inappropriate choice of defaults. An unfortunate choice for representing missing values is 0 with no way of telling a missing value 0 from a genuine 0. A revenue manager was once shocked at how many customers were "zero-billers" until he figured out that over 60% of the 0's were missing values populated subsequently by a process that was delayed.

  Yet another source of data mangling are measuring devices with hidden limitations that can introduce potential DQ errors. Some counters reset themselves after they reach a maximum. Some devices cannot measure more than a certain quantity. Some measurements are too insignificant from a revenue perspective and are dropped all together. These introduce **truncation** (some observations are dropped and the data set is truncated) and **censoring** (some observations are incomplete) errors that we will discuss in Section 5.2.2.

- **Data Loss**: Data can be lost for many reasons and in many ways. Just a few attributes could be missing or entire records and files can be lost. Buffer overflows (e.g., running out of disk space, overflowing pre-allocated memory) are a major reason for data loss and unexpected behavior of software that can lead to faulty data. Similarly, during transmission, if either the receiving end or the transmitting end have problems, files can be lost in transmission. Processes are often not properly designed to perform checks and to ask for re-transmission of files should something be missing.

Potential Solutions

The solutions for the DQ problems that arise during data delivery are predominantly process based.

- **Transmission Protocol**: The transmission protocol that is used for data exchange should permit the running of validation tasks to confirm that the entire data has been transmitted. There are many protocols with different capabilities, such as CONNECT:Direct and FTP. Each protocol has different functionalities and is compatible with different operating systems.
  - **Relay Data**: One way of avoiding data loss is to relay data to intermediate sites. This way, the transmitting server doesn't get backed up if the receiving server is unable to receive the data for a short period.
- **Verification**: Run verification tasks to make sure that all the data have arrived, and that there are no spurious data. The sooner that problems are detected, the better the chance of making a successful recovery.
  - Typically, there is a predictable pattern in the set of received data files. For example, on Wednesdays you would expect 20 files from the New York office totalling 180 Mbytes and 4 files from the New Jersey office totalling 7 Mbytes. The file names will often indicate the contents (e.g., NYC_June_3_2002.dat), and often there is a data file header containing additional identifying information. The patterns in file delivery can be readily discovered using the EDM techniques discussed in Chapters 2 and 3. Raise an alert if the set of received files significantly deviates from the expected set.
  - Verify that the files are in their expected format, and contain the expected data. Consider writing a lightweight version of the file parser which verifies the file format, and that for example the

NYC_June_3_2002.dat file contains data for the New York office and not the Boston office.

- **Feed Integrity**: Ensure that the feeds and files that constitute the data do not have integrity problems (perhaps requiring a more intensive processing than the transmission verification step) before they are combined with other feeds. Ask for *retransmission* if there is a discrepancy between what has been sent and received.
- **Feed Format**: Make sure that there is enough documentation (record format, file format) to be able to interpret the data and its structure.
- **Relationships**: Most data processes have interdependent feeds that need to be understood and synchronized to create the complete data. While delivering these feeds it is important to verify the inter-relationships among feeds and the relationship between the feeds and the source. Some questions to ask are:
  - Is the feed that we get an incremental view (changes since the last view) or a replacement? If it is incremental, do we have the baseline to apply it to? Should we obtain a replacement from time to time to make sure that the two views are the same? Very often, the views get out of sync quickly due to the fact that the rules used to apply the increment to get a current view are seldom completely documented. New rules ("if the value is 'NULL' change it to 0 because our software doesn't like mixed mode") are made spontaneously to force a recalcitrant process to run. The downstream user never finds out until the reconstructed view gets out of sync with the original.
  - Are the data coming in a stream ("packets that are flowing through the router") or as a database dump (census data)?
  - Do we know the dependencies between the feeds? Which feed needs input from others for processing?
- **Interface Agreements**: As we have seen earlier, data are unusable if we cannot understand and interpret them. It might be worthwhile negotiating a contractual agreement wherever possible to have access to the (a) data dictionary and other documentation (b) notification whenever there is a change or update to the process and (c) a well-established mechanism for resolving problems.

We note that the most of the data delivery solutions are process related, including accounting for all the data and making agreements to get information (data dictionary, business rules), to understand and interpret the data (metadata).

As seen in the preceding discussion, an important part of process based data quality control is closely monitoring the data flows, checking at various stages for data leakage and data publishing. We discuss below *data monitoring* and *data publishing*, which should be incorporated in the data flow to improve the data quality.

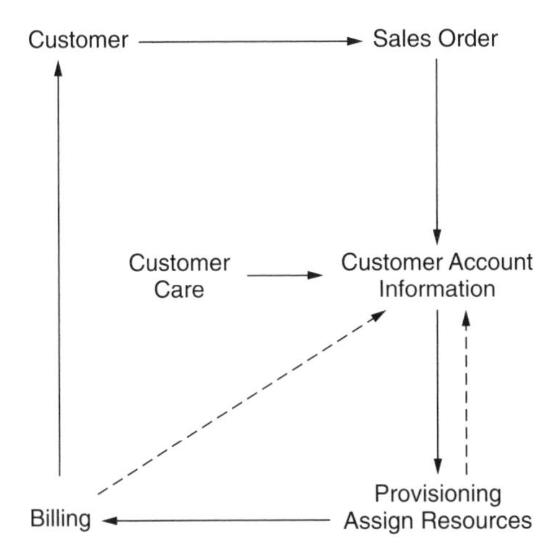

Figure 4.2: Feedback loop in a process.

4.3.3 Data Monitoring

Let us consider how a new transaction (sale, Internet session, phone call) originates and flows through the work flow process being measured and recorded in various databases on the way. We use the highly simplified example of a process flow in Figure 4.2. Typically, every stage is represented by a database or a collection of databases that contain every transaction or action taken by the process. For example, a customer's request for a service or product will result in an input to the database. The information is passed on to a sales order system that will create a request and also fill in the customer specific information such as name, address, and request date. The information is then passed on to a provisioning group that fulfills the order by arranging for various resources (a port, an installer) to complete the order. The specific resources allocated are recorded in a database. Finally, the customer is billed for the services. There are recurring and nonrecurring charges that contribute to the bill. Each of the stages in the data flow is represented by a complex set of databases and data warehouses.

In addition to the internal consistency of these databases, problems arise during handoffs between databases. We have found this to be a recurring problem. At a very fundamental level, disagreements between definitions of data constraints and business rules cause significant data quality issues. Sometimes, the owners of the process are not even aware that such inconsistences and differing implementation of the same data and business rules exist.

Data quality problems in workflow data have dire consequences such as:

- **Revenue Loss and Revenue Assurance**: A significant percentage of the transactions do not make it from the databases that measure the usage to the billing systems. It is estimated that in the telecommunication sector, on average between 3% to 8% of total revenues are lost, often reaching as high as 15%. **Revenue assurance**, the field of plugging revenue leakage by accurately billing all usage, is closely tied to data quality improvement.

- **Accurate view of resources**: From the perspective of provisioning, it is important to have an accurate view of the resources available, whether it is the telecommunication network or a cineplex or airline bookings. If a resource is not de-allocated when the customer leaves, the company will either have to deny service that it can actually deliver or spend precious cash obtaining additional resources.
- **Inability to provide new services**: If the customers requests online access to manage their own resource needs in real time, a corporation that does not maintain accurate data processes will not be able to offer it, losing valuable business to competitors in the process.

In addition to technical and process related reasons, there are many sociological reasons that can be obstructions to smooth data flows.

- **Organizational Boundaries**: Organizations seldom share information, even though they are all a part of the same data flow process. Each organization is focused on improving data quality within. However, if the systems have interdependencies, it is not sufficient to clean up within an organization. In fact, locally improving a database might have the opposite effect on other databases further downstream by making them incompatible.
- **Transitions**: Information, typically metadata that relates to conventions about processing the data and hidden rules, are not communicated properly during personnel, project or company transitions and are lost over time, making the data opaque, riddled with hidden pitfalls.

For all of the above reasons and many more, data monitoring is a critical step in data management and data quality.

Methods for Monitoring Data

Data monitoring is a continuous process as the name suggests, since data quality issues arise constantly and cannot be solved in a single one-shot clean up initiative. An incomplete list of solutions is:

- **Track Data**: A simple method of tracking data is to follow instances as they flow through the process. Sampling vertically at any given point in the process will not help. Furthermore, the longitudinal tracking of data has to be done regularly since problems arise over time. Continuous or online auditing systems are being developed to address the need for data monitoring continuously in time.

- **Reconcile and Validate**: When the feeds consist of incremental updates, it is easy for them to get out of step with the original feed due to missing information. It is good to validate the data by **reconciling** it with the original periodically.
- **Mandate**: A corporation wide diktat could potentially mandate that a minimum set of data elements should be maintained or obtained from a **database of record** to ensure that the data are consistent across the corporation. Pressure from upper management is an extremely effective tool in obtaining DQ compliance.
- **Feedback Loops**: Perhaps the most important data monitoring tool is the **feedback loop**. There are natural feedback loops in some instances. For example, overbilled customers will usually call the company to complain. However, this process is asymmetric—most customers will not call to report underbilling (or not being billed at all). Therefore, a company needs to put in place its own feedback loop to check for errors. In Figure 4.2 the dotted arrows represent loops that need to be in place to make sure that the databases of customer accounts, provisioning and billing are always able to communicate with each other.

  Whenever a change is made anywhere in the system, the changes have to be propagated and cycled throughout the system to ensure that every system which uses that data element is informed. At the very least the data base of record (which serves as the ultimate authority for that piece of data) should be updated. If a customer calls to disconnect a particular service or product and the provisioning system is not informed, the resource will not be deallocated and soon all the resources will be flagged as "in use" when they really are not.

Data Publishing

As an important data monitoring strategy, data are made accessible to the users through **data publishing**. The motivation for data publishing is often the scale of the data. The raw data are overwhelming in size and complexity, and integrated from several sources. The task is made easier for the end user of the data by preprocessing the data and making it available in a manageable form through publishing. The access to the data is often provided over the web. Data publishing also involves publishing a *description* of the data serving as documentation. Data integrity problems and their resolution are also posted serving as an evolutionary history. Most of all, data publishing allows many users to scrutinize the data, each using a different portion of it, in a different way, for a different purpose. This serves as a "testing" phase for the data since the users are sure to report data problems, increasing the likelihood of detecting data quality issues. However, a drawback is that the data problems pointed out by the users could be potentially used to discredit the data source and used as political ammunition.

In order to publish data, massive data are summarized to manageable proportions, using *data squashing* techniques among others. In addition to summaries, publishing exceptions serves as a **data alert** mechanism.

Data Squashing

Data squashing is an important example of the use of EDM for data quality. Three important EDM techniques for squashing data are:

1. **Aggregates based on data cubes**: We described data cubes in Section 3.2. A common form of squashing involves computing aggregates such as frequency counts, sums and other "summable" aggregates.
2. **Parametric Data Squashing**: Dumouchel et al. in their 1999 paper on data squashing propose using parametric methods to impute values within a cell of a partition using various distributional techniques such as the multivariate Gaussian. With such an assumption, we can compute any aggregate that is required.
3. **Nonparametric Data Squashing**: DataSpheres are the basis for a nonparametric squashing of data. The boundaries and the summaries are dictated by the data and are mostly assumption free.

4.3.4 Data Storage

Once the data have been delivered, they must to be stored on devices in some format and with some storage management software. In these days when a corporate operations database can easily fit in an inexpensive desktop PC, physical storage is generally not a problem. We do not address physical storage in this book except to note the problems which can occur (e.g., insufficient space, limits on file size, data loss).

This book is more concerned with the logical storage of the data—whether in a collection of files, in a database, or in some other structure. In any case, a schema must be developed, storage conventions decided upon, and metadata developed for the end user. The major DQ problems that arise during storage are:

- **Lack of Awareness and Planning**: Decisions are based on outdated notions of storage and capabilities. It is possible to store terabytes of data on PCs now at very little cost. However, people continue to take shortcuts in planning and storage to save disk space.
- **Paucity of metadata**: Metadata are data about the data that we use to understand and interpret the data. Typically metadata are stored as documents and data dictionaries. They tend to be incomplete for several reasons. Often the rules and exceptions that govern corporate data are so complex that they are not documented completely. Even if they start off complete, many changes are made along the way that are not added to

the metadata store because of time pressures or simple oversight. In legacy systems, valuable metadata are lost due to personnel changes. They often exist only as oral traditions passed on from administrator to administrator, seldom shared with downstream users. In such situations the user might have to "discover" the metadata by browsing the data. We revisit this topic in Sections 5.3.5 and 5.4.

A common problem is that the metadata and data exist independently of each other, even though most DBMS provide facilities for embedding the metadata within the DBMS. However, these facilities are not always used and only the minimum amount of data description is documented.

- **Inappropriate data models**: It is essential to represent a data element with an appropriate data structure. A common problem is a failure to properly represent historical or time series information, because only a current snapshot is stored. For example, an employee might receive a raise in July. If a database stores only current salary, it cannot accurately compute the amount the employee is paid in 2002. A proper structure records not only a salary, but also the time period during which that salary is valid. Such mistakes arise when people with inadequate skills design the data storage. Therefore, it is important to pick the people with right skills to handle various aspects of the data flow.

- **Ad hoc modifications and changes**: Decisions are sometimes made in a myopic fashion, due to time pressures. For instance, a particular GUI (graphics user interface) might allow only eight characters in an input field so a decision is made to report a particular revenue in dollars rather than cents. If this standard is not implemented for all data centers and the data users made aware of it, DQ problems will arise. When the data are integrated for a global analysis, misinterpretation such as the one in the example at the beginning of this chapter (T. Das's revenue was 24, Ted J.'s was 2000) will occur.

- **Software and hardware constraints**: In the previous example, the limitation of the GUI capabilities introduced a DQ error. Similarly, other measuring devices introduce DQ errors as well. For example, a metering device might be able to measure calls only for 3600 minutes, and will not record anything higher. This introduces **censoring** errors where we can only observe a part of the data value. In some other cases, due to limitations on disk space, certain items are not recorded at all. For example, calls less than 2 seconds might not be recorded. This procedure alters the number of data points and is called **truncation** since the data set gets truncated. Censored and truncated data are also called **incomplete data**. By contrast, **missing data** can occur when storage space runs out. This problem can happen when a file system fills up, or because of constraints on the maximum size of a file (2 Gbytes on many systems).

  Software and hardware constraints can lead to other forms of data mangling in a short-sighted effort to save a few bytes. The most famous example is the Y2K problem, but the problem occurs in many forms, For example, using too few digits in a "unique identifier", so that the identifier numbers must be recycled (and therefore perhaps duplicated).

Potential Solutions

The DQ problems that arise during data storage have reasonably actionable solutions. Most of the solution lies in planning ahead and keeping information updated.

- **Metadata, metadata, metadata**: Clear and accurate information about the data is the biggest DQ problem encountered in practice. It is also the hardest to solve, since the analyst who is the user of metadata has no control over how and how much of metadata are documented. Mandating accurate and updated metadata and rewarding compliance are an important solution to DQ problems. In the next section we will discuss technologies that are available to facilitate managing metadata.
- **Plan Ahead**: Many DQ problems arise when there is a mismatch between the hardware, software, expertise required to handle the data, and the actual resources used. Planning ahead to size the problem, project future needs, hiring the right skills, and building in appropriate checks and controls can help mitigate the problem. DBMS have the ability to build-in constraints that can automatically check to see whether data elements meet the specifications defined by the constraints. We will discuss this in the next chapter.
- **Data browsing**: Exploring the data is an effective way of discovering hidden rules about the data as well as unearthing obvious flaws such as missing chunks of data, censored data, and truncated data. There are several tools available for interactive data browsing, both broad data mining oriented products (SAS, IBM data mining products, Evoke data profiling) as well as smaller cutting-edge research efforts like Bellman, devised specifically for data quality and detecting hidden relationships in data.

4.3.5 Data Integration

The most intractable DQ problems arise during data integration, the process by which multiple data feeds are brought together to create a rich and complete dataset. Most corporations are creating **federated data**, where data from different sources are stitched together to form a whole. For example, a company might have a unit that sells detergents and another that sells cosmetics. It would be reasonable (provided no customer privacy rules are breached) to bring together the data from the two units to study associations between detergent and cosmetic buying patterns. On other occasions, companies merge or acquire each other. If they have a common customer base, it makes sense to combine the data. The major DQ challenges during data integration are:

- **Multiple Data Sources**: When data from multiple sources have to be integrated, we need a common **match key** or **join key** to join the data together. For example, if we know the social security number, we can bring together (hypothetically) the credit history and medical history of an individual. The lack of a good key is a common DQ problem.

  In the absence of a definitive key like the social security number, integrators resort to **approximate joins** based on messy fields like names and addresses that have no standardized representation. Or, if there is a declared key, say a phone number, but the match rate is not good, analysts resort to *arbitrary matching heuristics* such as "match on key1 from dataset A and key2 from dataset B, if that doesn't result in a good match rate, also try matching key3 in dataset A and key4 and key5 from dataset B, because they look like phone numbers". This can be considered an **inferred join**. While this might improve the match rate, who knows what the resulting data represent?
- **Different Definitions**: A related problem is that different organizations in a company have necessarily different views of the same entity. For example, a customer could be a contract, a service order, a bill, an assigned physical resource, or a designated site. Each organization would maintain its own version and the mappings could potentially be many-to-many. In such situations, integrating the data requires a careful understanding of the underlying meaning of a "customer" and the data that are used to represent the customer.
- **Time synchronization**: When dealing with multiple feeds, especially those that change frequently, it is important to see that they are synchronized properly during data integration. For example, if we are combining network traffic data and network performance data, it is important to get the feeds synchronized correctly, otherwise the analysis will produce misleading results.
- **Unconventional Data**: There are many new data types such as audio, video, images, web pages, web server logs, and text. Storing and combining these data presents new challenges that are not yet fully explored or understood.
- **Legacy Systems**: Aging legacy systems pose difficulties because of internal inconsistencies that have been compounded over the years. Furthermore, they function on outdated hardware and software platforms so that sometimes the data cannot be accessed. However, there are commercial firms that specialize in moving legacy systems onto more modern platforms.
- **Sociological Factors**: The biggest hurdles to data integration are sociological factors. Organizations within the same company do not like

sharing their data, either because it dilutes their power or it might bring to light any flaws. Sociological and political factors make obtaining data and metadata very time consuming.

Potential Solutions

- **Mandate accurate time stamps**: As mentioned earlier, certain precautions like insisting on accurate and detailed timestamps can help synchronize feeds. However, it might not be operationally easy to implement it.
- **Commercial vendors**: There are commercial vendors that will scrub, profile and migrate data. We do not want to mention any by name; however, a simple search on the Internet will bring up the major companies that provide services in this field.
- **Browsing Tools**: Exploratory data browsing can identify hidden relationships that can point to ways of integrating data by identifying simple and multi-part match keys. We will describe some techniques in Section 5.3.5.

4.3.6 Data Retrieval

Problems in data retrieval are primarily resource constraint and resource mismatch issues.

- **Human Error**: A frequently occurring but easily solvable problem is human error. However, human error problems cannot be solved on a large scale for massive data. A query is not properly written, the wrong attributes are specified, or there is some other error not related to syntax. Such errors can result in the wrong data being used for analysis (April bills analyzed instead of May bills) or the retrieval process taking an inordinately long time due to improperly structured queries.
- **Computational constraints**: On occasion, the data are overwhelming in size, so that in order to accommodate computational and space criteria we have to modify the retrieval request. It might take too long to retrieve minute-by-minute aggregates of network traffic for a two month period, so we decide to use only hourly aggregates. Such approximations impact the analysis further down the data chain.
- **Software incompatibility**: A frequently encountered problem in practice is that the data are provided in a format that the user cannot read. For example, the database dump is in a particular version of a DBMS software but we have some other incompatible version. The negotiations during data transfer often do not get down to this level of detail, and so data access is delayed.

Potential Solutions

The solutions to DQ problems arising in data integration are primarily process based.

- **Tools**: We should use appropriate tools such as XML for data interchange that will tie metadata to the data. Furthermore, data browsing and EDM tools (e.g. Bellman) help in discovering functional dependencies and join paths when the metadata are missing or inaccurate.
- **Testing**: The queries used to retrieve the data should be rigorously tested and the output verified to make sure that the data retrieved are the data we want to retrieve.
- **Planning Ahead**: Before retrieving the data, we should plan what kind of data, how much, when we need them and to choose the right tools to retrieve them from the format in which they are stored.

4.3.7 Data Mining/Analysis

Traditionally, the analysis phase is not considered a source of data quality issues. However, given that the ultimate result is a combination of the quality of the data as well as the quality of the analysis, we cannot neglect the analysis aspect, particularly in its suitability to the data.

$$DATA + ANALYSIS = RESULTS. \tag{4.1}$$

- **Scale, performance and confidence guarantees**: If the dataset is massive, the issues of scale and performance can introduce DQ problems. Certain techniques like classification and particular types of clustering algorithms are computationally expensive. In order to reduce the data to manageable size, analysts resort to sampling, that is, choosing a small subset of the data. While sampling works when we are interested in aggregate and typical results, it is not suitable if we are interested in peculiar or outlying data points. On the other hand, some methods perform well on large data sets but do not provide confidence guarantees. Fast clustering methods like k-means are an example. While they guarantee that the clusters are optimal according to a specific objective function (e.g. minimize within distance and maximize between distance), they provide no measurement (let alone guarantee) as to how well the model fits the data.
- **Black boxes and dart boards**: At one point in the mid 90's, it was believed that data mining could and would replace a human analyst. Raw data would go through a data mining "black box" and emerge as succinct, interesting patterns or results. However, most analysis involves domain-specific knowledge that cannot be incorporated into black boxes. A related approach is to throw many models at the data hoping that at least one would hit the target. However, this is wasteful and does not result in a high success rate. Worse, there might be a spurious, purely accidental match between the data and the model resulting in dangerously misleading conclusions. The resources are better spent in EDM to learn from the data the kind of models that are appropriate for representing the data.

- **Attachment to Models**: Analysts are known to get attached to particular types of models. Many criteria drive the choice of the model—simplicity, easy to understand and interpret, easy to implement. All these are important but the most important criterion should be the match between data and the model. This can be evaluated initially by using EDM methods to sequentially refine the model choice and using "goodness-of-fit" methods to determine the suitability of the models to the data.
- **Insufficient Domain Expertise**: Domain expertise (DE) is essential at every stage of the data flow process. What data need to be collected, what they mean, how they should be used and how the results should be interpreted as well as assessing the importance of the results all depend on DE. In many data mining exercises, the top 10 results are findings that can be ascribed to not taking into account hidden "business rules". While nothing can replace DE, browsing the data using EDM techniques can help us understand the characteristics of the data. For example, the revenue attribute is highly correlated with usage attribute but not with the length-of-tenure attribute.
- **"Casual empiricism"**: Casual empiricism is the use of numbers that are not backed by rigorous reasoning. This is particularly true of thresholds and cut-offs. Anything above 10 is "big", between 5 and 10 is "medium" and less than 5 is "small". It could be that 99% of the data lie in the "medium" category, making the thresholds almost meaningless. Such parameterizations are often done before exploring the data.

Potential Solutions

In the next chapter, we will discuss statistical techniques that can help address DQ concerns. We mention below a few general solutions to ensure DQ during the data mining stage.

- **EDM**: Exploring the data using EDM techniques (data browsing) is an excellent DQ strategy. It increases our understanding of the data, identifies peculiarities (long tailed distributions, appropriate variable transformations) and detects dependencies and associations among attributes for the purpose of selecting attributes for inclusion in models and discovering join paths. It helps us to discover certain types of "hidden" metadata and identify potential join paths between tables as well as databases. EDM enables us to hone a modeling strategy suitable for the data.
- **Accountability**: The analysts should be required to justify their choice of analysis and be accountable for the results, irrespective of whether they have moved organizations or not.
- **Continuous Analysis**: Data changes from day to day, either due to changes in the entities being measured or due to changes in the data processes. One-shot analyses do not capture this dynamic aspect. Therefore, we should design analyses to function continuously and reflect the changes.

- **Sampling Versus Full Analysis**: Sampling is used when the raw data are overwhelmingly large. Many techniques like neural networks and nonlinear techniques like logistic regression do not perform well on large data sets. However, sampling works well when we are interested in "typical" behavior or in events that happen frequently. Sampling will not work if we are interested in every individual data point or rarely occurring patterns.
- **Feedback loops**: Analysis can serve as a good DQ control check point. The findings of the analysis should be used to suggest improvements to the data gathering processes in terms of what, how much and how data are collected. We indicate this by a dotted line in Figure 4.1.

Let us revisit the definition of data quality in the context of the data quality continuum.

4.3.8 Annotated Bibliography

See [30] for a detailed account of data errors. The paper [14] discusses the role of data stewardship and its impact. In the context of data loss, see [87] for an introduction to buffer overflows. Please see [13] for an example of revenue assurance in the telecommunications industry. See [17] for a general discussion of continuous auditing.

Data squashing is discussed in [39] where a parametric approach is used. Nonlinear squashing is discussed in [71]. Bellman, a data browsing tool that profiles data, discovers functional dependencies and infers join paths is discussed in [33].

4.4 THE MEANING OF DATA QUALITY REVISITED

Conventional DQ metrics focus on the data meeting stringent specifications such as accuracy, completeness, and consistency. However, in the light of the discussion in the preceding section, the metrics are inadequate to cover all the aspects of data quality. We need to update and expand the list of DQ metrics to create a more comprehensive and practical definition of data quality.

In addition to the schema-related constraints that are generally applicable to any data set, there are constraints that are specific to an application, expressed as a set of business rules that should be reflected in the data. For example, a company's inventory might consist of some items that are for sale to the public and other items for its own use. The business rules that determine the nature of the items (red ones for internal use only) will be implemented in the data flow process (if color=red then use=internal; else use=for sale) and reflected in the database as:

item_id = 1233, color = red, use = internal
item_id = 1234, color = white, use = internal
item_id = 1235, color = blue, use = for sale

Notice that item 1234 violates the business rule. Both types of constraints play an important role in determining data quality.

4.4.1 Data Interpretation

It is essential to have the correct and complete information to interpret data (metadata). This might be in the form of a data dictionary, a schema, or a set of rules to apply to the data. Some rules can be quite complex. For example, "if attribute A has value *N*, then generate 20 additional records, with the value of attribute A going from 1 to 20, all other attributes remain exactly the same." The data are meaningless if this rule is not made available to the data user. Such rules are frequently used in practice for informal "data compression". The data models have to be specified clearly and updated when any changes are made. We should check to see that the data conforms to the models specified. Finally, the data and metadata have to be easily accessible. The above criteria measure the **interpretability** of the data. Low interpretability is a significant source of DQ issues.

4.4.2 Data Suitability

DQ problems arise when a data set is not suitable for the question being posed. For example, we cannot "analyze" the performance of the network if all we have are configuration data with no traffic data. While this is an egregious example, there are practical instances where snapshot data have been used to analyze behavior that changes with time. Therefore, it is important to determine if the dataset contains *relevant* and *sufficient* information to answer the questions we pose. Similarly, it is important to determine whether the analysis options available are suitable for use on the data.

4.4.3 Dataset Type

The DQ metrics, the definition of DQ, as well as the tools used depend on the nature of the data set. Each has a different set of challenges and peculiarities.

Federated Data
Primary challenges in enterprise data that are woven together from different sources are those of *sparseness* and *improper joins*. The DQ metrics should take into consideration the fact that when data are "missing" it is probably due to the fact that attribute is not relevant rather than it being measurable but missing. For example, not every customer subscribes to all the services a

company offers. When an integrated view of the customer is constructed, attributes corresponding to unsubscribed services will not be relevant and should not be considered as missing. Therefore, the conventional metric of "completeness" has to be updated in this context.

Federated data require that disparate data sources be combined. When no clear way to match the datasets exists (which is almost always the case) approximate matches are made, based on fuzzy criteria. This could lead to improperly matched records. For example, consider:

$$T.Das|9733608000|20calls|54minutes|Billed06/01 \tag{4.2}$$

and

$$TedJ.|9733608000|350calls|1000minutes|Billed07/01 \tag{4.3}$$

For example, the above two records might be "matched" based on the phone number, however additional data (not seen here) would reveal that the phone number was recycled from T. Das to Ted J. before the mandatory wait period of 60 days was over. This happens frequently in metro areas where there is a shortage of phone numbers. In the absence of additional data, we would infer that the customer T. Das has seen a sudden surge in usage. The attribute name should have been used to validate the match but this is seldom done in practice. We will talk more about validation of joins in Sectiom 5.3.4.

Massive, High-Dimensional Data

Scalability of techniques and processes should be a significant component of the DQ metric while considering massive data. Failure to do so would result in backlogs, processes exiting without completing, and so on, giving rise to all kinds of hidden and unpredictable errors.

Descriptive Data

Descriptive data usually consist of many tables with complicated interrelationships. Typically, such data describe relationships, allocations, and assignments. Consider the following simple example which depicts allocation of resources to projects in a consulting company:

Company Id = 123456789
Business Unit = A
Department = 1
Department Manager = Ms. X
Member = Mr. Y
Member Expertise = Statistics
Member Assignment = 70% Project A, 30% Project B
Member Status = Fully allocated

To represent these data company wide, we would need multiple tables, each containing the details of a particular aspect such as Department, Project, and so on. When a new project is proposed, the database would reflect the availability of resources (personnel) with the requisite expertise for the project. It is important that the database should be updated whenever there is a change and the change should be propagated throughout the system. Otherwise, the company could turn down lucrative project contracts under the false impression that no staff is available, if the members are not "de-allocated" from the projects on completion. Conversely, if the assignment of members to project was not updated, the company could take on projects that it does not have the resources to fulfill. Real-life network configuration databases and other descriptive databases are incredibly complex and difficult to validate ("Is customer A really homed on port 1 in cabinet 2 in building A on Main St, Suburmania, NJ?"). DQ metrics and solutions are primarily process based—enforcing strict feedback loops, continuous auditing, and validation of random samples.

Longitudinal Data

Time series data are important in the study of life cycles, periodicities, trends over time and for forecasting future behavior. From a process perspective, we can approach longitudinal data in two ways, either by observing the value of an attribute at a given point in the process as it changes over time or by following an individual data point over time as it passes through various stages of the process. From a DQ perspective, both approaches help us identify problems—process flows, the way in which data tend to get corrupted over time and the processes that contribute to it. We can infer propagation of errors and their compounding as they pass through the processing stages. In the absence of metadata, longitudinal studies can help us learn about the process flows.

An important DQ component of time series data is **synchronization**. Misleading results and DQ errors happen when time series are correlated improperly. For instance, in a study pertaining to inventory flows, an initial analysis revealed that a large proportions of components had conflicting labels when the snapshots of three major databases in the end-to-end process were compared. The analysis had accounted for the time taken for the data to flow from one database to the next. However, the time lags in reality did not match up with the lags that were documented and used by the analysis. As a result, while DB-A and DB-B had fresh data, DB-C still had the previous week's view of the data, resulting in over 5% of the data (thousands of records) being rejected as flawed. The inventory case study is described in detail in the next chapter, as well as a case study in using statistical techniques to identify DQ problems in time series data.

Streaming Data

A **data stream** is a sequence of data emanating sequentially at a high rate of accumulation from a single source that we get to observe (for practical purposes) just once. In real life, data streams arise in telecommunications (call detail records), network related studies (performance data derived from polling routers) and meteorology (measurements on weather-related phenomena, particularly fast-changing ones such as storms and cyclones). The challenge is to devise analyses that take into account the volume and rate of data accumulation as well as the fact that we do not get to see all the data at any one given time. Data streams require **real-time DQ**.

Here is an example of a data stream where the columns in order are time-stamp, calendar date, time, and four types of network traffic measurements:

```
948205525 01/18/2000 09:25:25 1379358875 1767967974 188471501 1917803055
948205585 01/18/2000 09:26:25 1379597278 1769003702 189510384 1917949117
948205645 01/18/2000 09:27:25 1379823191 1769738866 190248893 1918084438
```

Such data are plagued by all kinds of problems—lost data, counters resetting themselves, synchronizing multiple sources and so on. We need careful DQ consideration in the analysis to account for such problems. Furthermore, an additional DQ concern is the stability of estimates that are used to summarize and publish streaming data.

Web Data—Text Mining

Data that are "scraped" off the web are used frequently to leverage the abundance of data available over the web. Sometimes, the web is the primary or the only electronic source of particular types of data where they are published specifically for sharing, in order to comply with legal or corporate requirements. However such data are inherently messy as the data were not intended for integration with other resources.

Web-scraped data might be generated by collecting a set of web pages related to a particular topic using a search engine. The data set can be expanded by further collecting all pages within k links of the above collection of pages. Formatting tools (e.g., perl scripts) are then used to transform the data, for instance from HTML tables to comma separated values (CSV).

Data scraped from the web have unconventional formats and low DQ standards. Sometimes, the data are intentionally corrupted so that 212-555-1212 is misrepresented as 2i2-555-i2i2 or name@domain.com might be represented as "my last name at domain dot com" (to avoid being picked up by programs that harvest information from the web).

Another flavor of web data are web server logs. When a user sends a request to a server, a log is generated which contains a wealth of information. The log entries can be written in various formats, one of which is the *Common Log Format (CLF)*. This is a standard format that can be produced by many different web servers and read by many log analysis programs. We have taken the following example from the Apache web site http://www.apache.com.

127.0.0.1 - frank [10/Oct/2000:13:55:36 -0700] "GET /apache_pb.gif HTTP/1.0" 200 2326

This line has several parsable components, for example, that the user "frank" at IP address 127.0.0.1 issued a "GET /apache_pb.gif HTTP/1.0" request at [10/Oct/2000:13:55:36 -0700]. Some pieces of information are missing (e.g., the "-" indicates that RFC 1413 identity of the client is not known), and other pieces should be viewed with suspicion (e.g., is 127.0.0.1 the actual IP address of the requestor, or is it the IP address of a proxy?).

Given the nature of web data as well as the eclectic uses it is put to, defining metrics and finding tools for cleaning such data are highly domain specific. Data cleaning involves scrubbing headers, searching for embedded text and other techniques. Apache as well as Yahoo and other major portals and search engines offer tools for analyzing and cleaning server log data. Such tools are very popular for monitoring security of web sites to prevent hacking and denial-of-service attacks.

Text data and its analysis are inevitably tied to information retrieval, natural language models, and computational linguistics, all of which are outside the scope of this book. There has been great focus and activity in these areas with the development of search engines like Google. However, in terms of data quality techniques, there are no well-defined techniques or references.

In the rest of the book, we only consider text data that occur as a part of a larger heterogeneous data set (character attributes), not text data as defined by a document of text alone.

4.4.4 Attribute Type

Just as the DQ challenges vary depending on the data set type, the DQ metrics as well as the tools used depend on the attribute type. Statistical methods like outlier detection, control charts and goodness-of-fit metrics can be used for evaluating the quality of numeric attributes in a data set.

As we saw earlier, descriptive attributes require more process-related, audit-based techniques. Such data are hard to validate since they cannot be inferred from patterns or trends the way numeric and even text data can be. Descriptive data might require a *sampling-based* approach where selected data are verified. We might choose a few hundred records out of thousands and have them validated by humans—"Is this customer's connection homed at this particular physical location? Is equipment A really located at the geographical address mentioned in the database?"

Similarly, character attributes would require tools that can match every value with a set of allowable values to verify the quality. For instance, if there is an attribute called Color which has permitted values red, white and blue, then a data quality validator will have to be able to scrub a value of "rde" and map it to the correct value "red". Such scrubbing is done frequently in data pertaining to names and addresses. We will mention a subset of such applications and techniques in the next chapter.

Web data and related text data require techniques that can recognize and work around pitfalls such as replacing 0s with os in phone numbers. Such techniques are so context specific that they often consist of perl and cgi scripts written by individuals on an as-needed basis.

4.4.5 Application Type

A DQ metric is closely tied to the application. Users have a high tolerance to missing or corrupt data if they are interested in aggregates and typical behavior. It is likely that they can get reliable results from a small portion of the data provided they can separate the bad data from the good data. They might resort to sampling, which can be a good strategy if the proportion of bad data is very small. Such a user might not be too concerned with completeness, focusing instead on the accuracy, accessibility and interpretability of the available data. On the other hand, users who need to focus on individual observations or abnormal observations in the tails of the distribution need to ensure a high degree of completeness.

4.4.6 Data Quality—A Many Splendored Thing

The preceding discussion shows that data quality is a very complex beast that cannot be contained with a simple clean solution that can be applied uniformly and universally. Overall, there are two broad components to DQ: A general component that is applicable to all datasets and a domain-specific aspect that varies from context to context. The general component consists of matching data and data models with specifications and constraints and looking for obvious inconsistencies in the data. However, the specification of the rules, constraints and the metadata that are needed for matching data against specifications are highly domain specific, with a "long tailed" distribution. That is, while there are a few rules that can specify and are applicable to say 50% of the data, subsequent rules are applicable to smaller and smaller chunks of the data so that we would end up with hundreds of rules just to specify 80% to 90% of the data. In other words, some specifications are almost on a case-by-case basis. For example, suppose that we are investigating data quality issues in a process that involves flow of consumer data from DB-A through DB-B to DB-C.

Rule 1: If Billing status = 1, then flow to DB-B and DB-C.

Rule 2: If Revenue lies in interval [A-B] then do not flow beyond DB-A.

Rules 3–10: Combinations of flows depends on combinations of billing status and revenue buckets. (At this point we would have covered say 60% of the data.)

Rule 11: If Billing status = 10, AND revenue greater than C AND state = NJ AND date between 2001 and 2002 AND company not equal Bad Co. then flow to DB-B but not DB-C. Rule 11 would cover an additional 1% of the data. And so on until.

Rule 256: If Billing Status = 12, revenue less than Z, state = TX, company = Oil Co. then flow to DB-C but not DB-B, which would provide a coverage of 0.0000001 proportion of the data.

Note that there are two intertwined issues here. First, in order to define the rules, we need access to top-notch experts. Second, the rules do not generalize to other data sets so that every exercise in DQ has this inevitable, highly manual and time-consuming task of developing domain rules that are necessary for validating the data. Answering this problem would require the interaction between several disciplines such as statistics, AI (expert systems, knowledge representation and engineering), process engineering and computer science, to name a few.

4.4.7 Annotated Bibliography

Please see [130] for an interesting application of mining data streams and for references to other literature in this area. See [21] for a discussion of integrating and using data gathered from the web. An example of the analysis of web server logs can be found in [70]. We refer the reader to a general tutorial on text mining [93] and to serve as a source for further references. An introduction to information retrieval can be found in [119].

4.5 MEASURING DATA QUALITY

By now it is clear that DQ is not a well-defined concept that can be measured and tracked with a crisp set of numbers. There are many subjective aspects. Some data glitches can remain hidden if they pertain to very rarely used applications of the data. For example, the information on T. Das might look like

T. Das|666-666-6666|666|66.66|-|666

which is clearly suspicious. However, checks on individual attributes or within aggregates would not have detected this peculiarity. It is likely that this glitch would have remained undetected until a specific query asked for T. Das. In other instances, there is just no way to determine if a data quality problem exists. How do we know if the "age" or "income" reported in a survey are correct? Therefore an important question to ask is:

Is the data quality component *measurable?*

At the outset, we need to establish the data and methods that are available for measuring DQ. In this section, we discuss criteria for measuring data quality and potential ways of measuring data quality. We revisit the topic in Section 5.5, and discuss the specifics in the context of a case study.

4.5.1 DQ Components and Their Measurement

The device for measuring data quality that we have found to be useful is to establish a set of *constraints* that the data should meet. We can then establish metrics by measuring how well the data meets the constraints. **Static constraints** (also known as schema constraints) are properties of the data itself—for example that a field is populated, that a "key" is unique, that two tables can be joined, and so on. **Dynamic constraints** are related to the end-to-end flow of the data through the process, as specified by *business rules*. An example of a dynamic constraint would be that a service termination is followed by a release of equipment within three days. Constraints give us a way to measure properties which are not immediately visible, especially properties of the data itself.

A way to classify metrics is to specify whether they are *operational* or *diagnostic*. Operational metrics measure our ability to achieve tasks using the data—they tend to be process related. While the end goal of a data quality improvement study is to improve at least some operational metrics, they are often too high-level to indicate what is wrong with the data. For example, "reduce the number of manual intervention by 10%" does not shed light on where the problem lies. Also, operational metrics do not indicate how well the data will perform for a new task. Similarly, it is not clear whether the improvement in the operational metric will actually improve the data quality in terms of reliability and usability for introducing a new e-provisioning service. Therefore we also use diagnostic metrics which are defined on the data itself.

In general, we will need to use all types of constraints (static vs. dynamic), and metrics (operational vs. diagnostic) to develop a usable data quality metric. We note that while we discuss a general list of criteria, the choice of DQ metrics and their actual measurement depends on the DQ implementer (see the case study in Section 5.5).

- **Extent of automation**: is a dynamic, operational DQ metric that measures the amount of manual intervention (hence the opportunities for introducing errors) required during the process. An approximate way to measure this is to *follow a sample of transactions* chosen randomly through the process and measure the number of human touches needed.
- **Successful completion of end-to-end process**: is a dynamic, operational metric. While it is difficult to measure whether a process conforms to specifications, we can measure the outcome. That is, what proportion of instances flow through the process with the desired outcome at the

end. For example, we might want to terminate transactions of customers with unpaid bills outstanding more than 60 days, while letting other transactions flow through. The sampling approach used above can be used here as well. The metric is the *proportion of instances in the randomly chosen sample that terminate with the intended outcome at the end of the process*. A more expensive alternative is to **simulate** the entire process, run various scenarios on it, and measure the performance. However, this requires a clear specification of the process which might not be available.

- **Glitches in Analysis**: is a static operational metric which measures the degree to which glitched data causes glitched analysis. We can measure analysis glitches by counting the number of findings (perhaps weighted by severity) which are invalidated or tainted by data quality problems. While analysis glitches can be difficult to measure, in our experience there is often some sort of feedback loop which accompanies the analysis, whether it is customers complaining about how their bill is calculated or analysts validating their findings before making a report.
- **Accessibility**: is a general diagnostic metric that is applicable to all DQ situations. One measure is *the time between request for access and the actual ability to view the data*. Another related measure is the number of *contacts* such as phone calls, e-mails, and so on, needed to get access to the data. Yet another measure is the *level of escalation* needed to get the access. A metric based on these aspects is suitable.

  As with any of these metrics, it is possible to cheat. For example, the keepers of a database might provide access to the data through a GUI with a predefined set of canned queries. The GUI might look good and be easy to use, measuring high on the "accessibility" component of a metric. However if the canned queries do not meet the needs of the user, then the metric is skewed.
- **Interpretability**: is a diagnostic metric that applies in a general context as well. The interpretability metric should be based on *availability of metadata* (e.g., counting the proportion of fields, tables, keys, foreign-key joins, and so on, which are documented), and the *adherence of data to specifications* (e.g., by counting the number of reported glitches that are resolved by updating the metadata).
- **Conformance to Business Rules**: is a dynamic diagnostic metric based on constraints of the type listed in Section 4.4.6.
- **Conformance to Schema**: is a static diagnostic metric which measures how well a snapshot of the data conforms to the metadata in its schema. For example, are the keys unique, do the values in the fields fit their formats, is there a single well-documented default value, are related fields joinable, and so on? While it is easy to generate a very long list of schema constraints, the data quality analyst must generally pick a select set of "important" constraints on which to focus.

- **"Traditional" metrics**: We have discussed the "traditional" metrics of accuracy, consistency, uniqueness, timeliness, and completeness in Section 4.2. While these are all desirable properties, we need a concrete way to measure them and to weight their relevance to the data's application.
  - **Accuracy**: measures the degree to which data values reflect the entity that they model, and can be regarded as a static diagnostic constraint. Unfortunately, accuracy can be difficult and expensive to measure. For example, we might want to establish the accuracy of a database which describes a warehouse inventory. Because accuracy measures a relationship between the database and the real world, we must incur a real-world cost to measure accuracy—namely, to hire a team to perform a warehouse inventory. Often, the data quality analyst resorts to sampling or the use of a proxy (e.g., complaints about accuracy) to lower measurement costs. However, the analyst must then be careful about introducing a bias in the measurement. On other occasions, as in a survey, there is simply no way to establish whether the respondent is telling the truth. A customer might say that they have no intention of switching to a competitor but might do so the very next day if an appropriate incentive ("we will give you $100 if you switch") or irritant ("Oops, we disconnected your service by mistake") comes along.
  - **Consistency**: measures the agreement between tables in a database or between databases. For example, a residential customer should be served by residential lines. Consistency is thus a readily measured static diagnostic metric (once we know which things should be in agreement and how).
  - **Uniqueness**: refers to the property that there is one record for each unique thing. A common way for duplicate records to enter a database is when the record's key is incorrectly entered, for example, the customer's name. *Duplicate elimination* (see Section 5.3.4) refers to the process of eliminating these duplicate records. We can measure uniqueness by counting the proportion of records which are removed. The caveats which apply to approximate joins apply here also—the "duplicates" might actually refer to different things.
  - **Timeliness**: What proportion of the data arrives as scheduled? What proportion of the data has accurate time stamps which will help us synchronize data feeds? What is the *lag between the origin of the data and its use*? How frequently do the data change? What is the ratio of the lag between updates to the frequency of change? If the data are updated in shorter time intervals than the intervals between data feeds, there is a serious danger of changes being overwritten, depending on how the process is set up.

    All such aspects have to be considered and weighed when designing a metric for measuring timeliness.

– **Completeness**: measures the degree to which we have recorded all relevant properties of all the things we want to record. Since completeness is a relationship between the database and the real world, it can be very difficult or even impossible to measure. In some cases completeness is merely expensive to measure (e.g., requiring an inventory of a warehouse, as with accuracy). In other cases, it is impossible (e.g., recording the identity of every Snark in the world). The data quality analyst frequently resorts to proxies to measure completeness, such as the number of null or default values in fields, the number of times a new thing unexpectedly turns up, and so on.

  We note that there is a flip side to completeness—the database should not contain entries for things which do not exist (such as disconnected customers or non-existent inventory). Fortunately, counting spurious records is usually not impossible, only difficult.

4.5.2 Combining DQ Metrics

We have seen that there are many measurements that can determine DQ. Can these be combined into a simpler set of numbers? We can consider a simple weighted average and determine the weights according to our needs—we can stress the dynamic, process related metrics more and down play the actual static, data related metrics if our purpose is to improve the processes. On the other hand, if we are more interested in the data itself, we might stress the static, data related metrics more. In our experience, a data quality improvement project generally starts by improving diagnostic metrics, then moves to improving operational metrics as the data becomes cleaner.

Ultimately, what is the purpose of the DQ metrics? It is to indicate whether the data are usable and reliable. In order to reflect this, an improvement in the metrics should be accompanied by an improvement in usability of the data and the reliability of the results obtained by analyzing the data. That is, the metrics should be **directionally correct**. Especially, diagnostic metrics should be directionally correct with respect to operational metrics. We illustrate this with a case study in Section 5.5, where we measure DQ by the increase in usability of the data and the increase in the automation of the data flow processes.

4.6 THE DQ PROCESS

Based on our discussions of the preceding sections we can outline a data quality process. Bear in mind that any such process has to be customized to the application as well as to evolving definitions and needs of data quality. We illustrate the DQ process with a case study in Chapter 5.

The DQ process starts with data gathered from different sources such as files, feeds, the web and so on. We can use various tools available for manag-

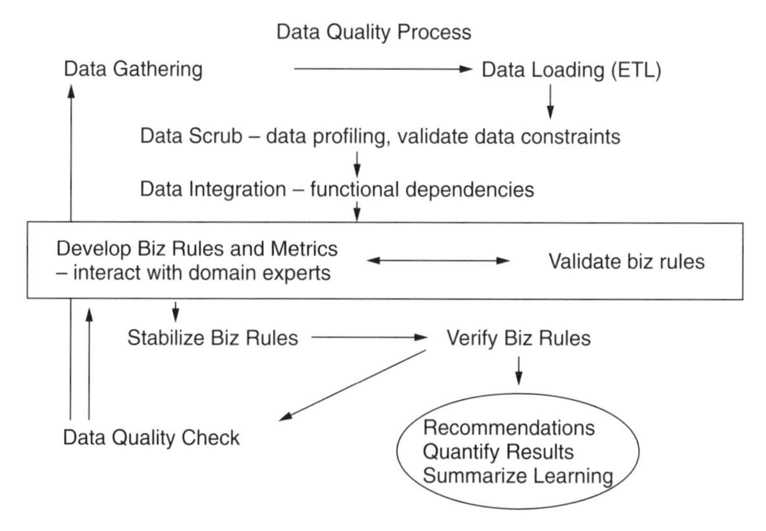

Figure 4.3: Data quality process.

ing these processes and protocols. The next stage of loading the data could use specialized ETL tools that we discuss in the next chapter. Data scrubbing and data integration involve schema constraint satisfaction and discovering functional dependencies for join paths to perform the integration. Data browsing tools are invaluable for this purpose.

The box in the middle is the most arduous stage—developing and stabilizing the business constraints and the problem specific DQ metrics. This stage is often the riskiest since the success of the project is heavily predicated on the success of this stage. It entails interaction with domain experts. In Chapter 5, the case study will illustrate the social and political issues that make this stage the trickiest.

Once the business rules are agreed upon, the process proceeds to validate the data against the rules and identify the faulty data for more rigorous inspection. The learning from this stage is fed back into the business rules and into the very beginning of the process.

The "good" data that satisfy the static (schema specific) constraints as well as the dynamic (business) constraints are certified and used for quantifying the quality of the data.

We have presented a very high level view of the process. As DQ practitioners, we need to employ a diverse arsenal of tools at each stage, whether mechanical, software related or good, old-fashioned people skills, to ensure success. In addition, note that there is a fair amount of customization of the process needed for each application (which we illustrate in Section 5.5).

4.7 CONCLUSION

We have seen that data quality is a complex concept with many aspects, some of which are measurable and some not. It requires intense interaction with the data as well as experts who can define the constraints that the data has to meet. In this section we give a brief description of four inter-related yet complementary approaches that have sought to provide solutions to aspects of data quality detection and assurance. These are drawn from process management, statistics, database management and metadata and domain expertise specification. We will give references to the process-related approaches having discussed the details in the context of the data quality continuum and definition of DQ. The remaining three approaches will be discussed in detail in the next chapter.

4.7.1 Four Complementary Approaches

We have touched on the various aspects of data quality that arise during the data flow process. Process-based management is critical given the complexity of the data flow process. We have seen that many DQ problems are born of bad process design and control and no data engineering for processes. The primary body work in the process management arena is in the TDQM program of MIT. The program mission is to "Establish a solid theoretical foundation for Total Data Quality Management, to devise practical methods for business and industry to improve data quality. Redesign business practices and implement new technologies in order to significantly improve the quality of corporate information."

In the past, statisticians had very few data quality problems. The problem was clearly defined a priori, experiments were designed carefully with a particular analysis in mind, and the data were collected meticulously. However, with more and more data being "found" data, statisticians have devised some techniques to deal with the flaws in such data sets. There are three broad categories of methods (a) missing and incomplete data, (b) suspicious data and (c) issues of the fit between data and analysis. Most data sets have a significant percentage of missing data, be it individual attributes or entire records. In federated data, missing data and sparse data are particularly egregious. If not addressed, it can lead to the wastage of 30%–70% of the data. There are many methods of imputing values to missing data, ranging from simple point estimates to complex regression and simulation based estimates. If the missing data are not addressed properly, the resulting analysis can be biased and misleading. Furthermore, tracking down missing data and studying their patterns can suggest solutions for DQ problems. On the other hand, suspicious data are those that deviate from what we "expect" the data to be. Such data are called outliers and we discuss methods for identifying and dealing with outliers in the next chapter. Goodness-of-fit methods can be used to investigate the suitability of the data to the analysis and vice versa, in keeping with our belief expressed in Equation 4.1 that data quality is a function of the data as well as the interplay between the data and the analysis.

The database community has developed numerous methods, prominent among these are techniques for cleaning up duplicates, correcting names and addresses and others. Other more recent approaches include exploring and profiling data, discovering functional dependencies among tables and attributes and using them to identify and validate match keys, finding complex join paths and others. We will discuss these techniques in detail in the next chapter.

Finally, the recording, transmitting and updating of metadata and domain expertise merit a thorough discussion given their critical role in data quality. There are technologies such as XML for the interchange of data that ensure that data and their definitions are closely tied together so that where there are data, there are metadata as well. Recording domain expertise is a more complex and manual process that might be helped by techniques from expert systems.

In the next chapter, we will discuss the above three approaches highlighting techniques and tools, and including case studies wherever possible.

4.7.2 Annotated Bibliography

Prominent in the body of work are [62] by Kuan-Tsae Huang, Yang W. Lee, Richard Y. Wang and [127]. A more comprehensive list of references can be found at the MIT TDQM web site http://web.mit.edu/tdqm/www/about.html.

CHAPTER 5

Data Quality: Techniques and Algorithms

5.1 INTRODUCTION

In the previous chapter we delved into the origin of many data quality issues, their impact, and the motivation for solving them. We emphasized the process because it has a direct bearing on the origin of a majority of data quality concerns.

In this chapter, we will focus on tools, techniques, and algorithms that can be used to address and repair some of the data quality problems that arise. However, there is no panacea, no single tool that can solve a majority of DQ problems. As mentioned earlier, data quality problems are highly complex and context dependent, requiring extensive domain knowledge and involving solutions that often need to be chosen case by case. In other words, a big part of the solution involves human involvement and expertise, so that complete automation is often not feasible. That said, automation can help with implementing checks and controls, isolating discrepancies, designing end-to-end audits that can be run at regular intervals and, in general, scaling up manual data quality checks to massive data.

We have seen that the DQ process involves verifying two distinct types of constraints (static and dynamic). The tools that we discuss in this chapter are drawn from several disciplines and are targeted at different aspects of the DQ process. These are representative solutions that can be customized to the problem at hand. *Data browsing* is a basic diagnostic activity—you cannot fix the problems until you know what they are. This activity entails many of the EDM methods discussed in Chapters 2 and 3 to get an understanding of the data set, as well as more specialized activities such as validating data definitions against their instantiation (validating schema-specific static constraints),

Exploratory Data Mining and Data Cleaning, by Tamraparni Dasu and Theodore Johnson
ISBN: 0-471-26851-8

cross-checking data flows end-to-end against business rules (validating dynamic constraints), summarizing typical trends (data publishing), identifying interrelationships, isolating outliers and analyzing them further. Other algorithms are used to detect and repair mangled data such as missing values, improbable outliers and values that have been incompletely recorded. Yet other methods can be used to integrate data sources. All these methods together help us to detect and fix to some extent the complex inconsistencies that are hidden in the data that data miners deal with on a regular basis.

In Section 5.2, we focus on techniques drawn from statistics such as missing value imputation, outlier detection and goodness-of-fit methods for detecting inconsistencies in interrelationships in attributes. In Section 5.3, we discuss tools specialized for use on database-resident data, such as Extraction-Transformation-Loading (ETL) tools, techniques for fuzzy joins, and advanced database profiling. In Section 5.4, we highlight the role of metadata and domain expertise in data quality implementation. The importance of metadata and domain expertise cannot be underestimated. The lack of metadata and domain expertise can make the data uninterpretable, frustrating any attempt at detecting or repairing data quality damage. In a majority of our case studies, a lack of metadata has been the most significant stumbling block. In Section 5.5, we present a case study of inventory building and revisit DQ metrics in this specific context. We conclude with a summary of data quality, its definition and measurement, and the accompanying challenges.

5.2 DQ TOOLS BASED ON STATISTICAL TECHNIQUES

Traditionally, statisticians dealt with data that was small and carefully collected. They could collect the data through careful design that was customized to the analysis technique which, in turn, was tailor-made to the substantive question, for instance, "Does the fertilizer A improve the yield of wheat? Is the effect of fertilizer A significantly different from that of fertilizer B?" In fact, many analyses such as ANOVA (analysis of variance) were actually developed in agricultural experiments where data such as crop yield, fertilizer amount, size of plots, and so on, were carefully measured. In other words, the statisticians were closely involved with the data gathering process and could control the quality of the data. Therefore, many of the techniques that we will discuss in this section were not developed with data quality in mind. They were developed to address small glitches (a missing measurement here and there) so that we need to be cautious when we use them on a large scale to address the systematic issues that seep into contemporary data sets.

There are three broad categories of techniques: (1) missing data, (2) incomplete data and (3) outliers. Missing data are a constant feature of massive data, where individual cells, columns, rows or entire sections of the data can be missing. There are techniques ranging from the naive to the complex for imputing values missing data. Sometimes, the data are not missing but *incomplete*.

For example, we might know that a process was still running after the 10-day observation period, but we wouldn't know exactly when it stopped. Therefore, all we know is that the system ran for at least 10 days. We will discuss other forms of incomplete data as well. On other occasions, datapoints are considerably out of line with what we expect them to be. Such datapoints, called outliers, are suspicious data points that could be potentially legitimate. There are several techniques for identifying, measuring and incorporating outliers into various analyses.

5.2.1 Missing Values

There are many reasons why there are holes in data sets. Federated data are prime candidates since they are formed by integrating different pieces which might have some common attributes, but also some that are specific to a dataset. For example, not every telecommunications customer has a cable modem. On other occasions, when the same exact data are gathered from different sources (e.g., sales from different branches of a retail store) a particular source might not send its data in time for the compilation of the integrated dataset. Yet in other cases, an individual sale might not be logged or some data element (e.g., customer phone number) might not be entered by the sales person. System and process related reasons abound as well. Another data quality issue arises when the same values are used to represent defaults and missing values. It is important to make the distinction between genuine zero-quantity bills and bills whose quantity have defaulted to zero because those data haven't come in yet for those particular bills.

Why Should We Care?

Why should we worry about missing data? First of all, we would waste an enormous amount of data (anywhere between 30% to 70% in federated data sets) if we threw away all data with a missing data field. Furthermore, studying patterns in missing data can help trace causes (e.g., all data from Trenton, NJ are missing) and reveal other DQ issues. Finally, missing data can introduce serious biases into the analysis since data seldom disappear in a random pattern. We might underreport network usage and loads if data from two or three high-end users are not reported. Justifiably, there is not a great deal of confidence placed in reports based on data with large missing portions.

Detecting Missing Data

Some types of missing data are easily detected. For example, we can *scan* rows and columns of data for gaps. We should *match* the data specifications against the actual data itself. Are there any attributes that are missing? If so, immediately check with the sources of the data. We should also *implement checks* during file transfer processes—check the number of files, the file sizes, the number of records and duplicates. Did we receive all the information that was sent? Another effective technique is to use *historical information*. We have data for a particular data point (e.g., a router) last n times the data was collected but we have no data this time around? What changed? While many of these checks seem simple, they are seldom implemented, resulting in serious DQ issues.

Other types of missing data are harder to detect. An effective way of detecting missing data are keeping track of *estimated* values and error bounds, such as total counts, averages, medians and standard errors. For example, we expect the new orders to be around 20,000 this month, give or take 1000. A considerable deviance from this estimate should trigger a more detailed analysis of the new orders. A particular concern in this approach is that the aggregates and estimates should be computed at a reasonable level of granularity (traffic per router, sales per metropolitan region), otherwise small discrepancies caused by missing data could be swamped in aggregates.

The problem of confounding defaults with missing values is more difficult to tackle. Access to historical data and domain expertise is needed. Domain expertise is needed to establish the existence of such a problem in the first place. Historical data would help establish the consistency of the value. If a value suddenly drops to zero and comes back to its previous value next time around, it is sure to be a missing data value that has been defaulted to zero. The **flip-flop** pattern is a good identifier of missing values in time series.

Occasionally, the pattern of missing data could lead to other data that could be potentially missing. For example, if spatial or geographic data for New Jersey are missing, it is worthwhile to check if all the data for Delaware have come in, because data for border areas might be missing even though there might be data recorded for Delaware as a whole.

Imputing Missing Values

The process of guessing the values of missing data is called **imputing missing values**. We should use the technique with great caution because while imputed values are good for aggregate analysis, no individual imputed value should be trusted (because it is an estimate).

The simplest imputation method is based on treating every attribute individually, ignoring any interrelationships with other attributes. Point estimates such as the mean or the median can be used to replace missing values. For example,

$$1,2,3,1,3,1,,,3$$

has three missing values all of which could be replaced by 2, the median of the non-missing values. Or, we could **simulate** a distribution using the non-missing values and draw from the distribution each time we encounter a missing value. So, in our simple example, the simulated distribution is:

$$P(1)=3/7, P(2)=1/7 \text{ and } P(3)=3/7,$$

so that the distribution of missing values will faithfully follow the overall distribution. Clearly, the assumption here is that the missing values follow the same distribution as the non-missing values. While the point estimate approach and the simulation approach are naive and based on potentially incorrect assumptions, they are very simple to implement, inexpensive to run and easy to understand and interpret.

We describe briefly more complex methods of imputation that exploit interrelationships between attributes and **impute multiple values** rather than a single value.

The multiple values reflect all plausible values for the missing value, rather than a single value. Imputing multiple values will result in multiple data sets (one for each of the multiple values). Each of the data sets are individually analyzed and the results are combined. A popular multiple imputation method is the *regression method*, where a regression model is fitted for every attribute using the previous attributes. There is an underlying assumption that attributes are **missing monotonically**, that is, in a data set with d attributes $Y_1, Y_2, \ldots, Y_d$, and if Y_j is not missing, then $Y_1, Y_2, \ldots, Y_{j-1}$ are all not missing. In other words, all the attributes preceding the attribute in question are populated for all data points. Monotonicity can be induced by using simulation methods described below. Here is an example of monotonically missing data.

Weight	Age	Height	Health_Index
20	2	10	5
15	5	9	3
25	5	10	.
20	4	.	.
10	1	.	.

The regression is preformed recursively, generating each attribute as we progress from left to right. So the first regression model would be:

$$Height = \alpha + \beta_1 Age + \beta_2 Weight.$$

After each regression, the missing values are replaced by the predicted value from the regression model plus an error term (a scaled normal deviate) and then the next regression model would be derived:

$$Health\_Index = \alpha + \beta_1 Age + \beta_2 Weight + \beta_3 Height,$$

and so on, until all the missing values are replaced with the regressed values. Since the error term varies randomly, we can generate multiple data sets by cycling through the imputation process. Each data set is analyzed individually and the analyses from the multiple data sets are combined to create a reliable single set of results.

Another method groups data by the propensity of any attribute to be missing, again with assumption of monotonically missing data. We describe the method briefly. An indicator variable (0 or 1 values) is built to indicate whether attribute Y_j is missing or not. A logistic regression model is built based on the values of attributes $Y_1, \ldots, Y_{j-1}$ to compute the probability that attribute Y_j is missing for any given observation. The observations are grouped by the probability that attribute Y_j is missing. Values of Y_j are generated for the missing values from the known values of Y_j within each propensity group using a mechanism called **approximate Bayesian bootstrap**, a discussion of which is outside the scope of this book.

For arbitrary missing value patterns, the MCMC (Markov Chain Monte Carlo) method is used to simulate data. The data are assumed to have a multivariate normal distribution. Next, the missing values are estimated by (a) building the conditional distribution of the missing values given the observed values and (b) computing the parameters of a multivariate normal distribution using the filled in sample. Steps (a) and (b) are repeated until the estimates become stable. The Markov chain refers to the tuples

$$(\tilde{Y}, \tilde{\theta}_i),$$

where $\tilde{Y}$ is the set of estimates of the missing values and $\tilde{\theta}$ the set of parameters estimated at the ith iteration. The values at the ith iteration depend only on values at the $(i-1)^{th}$ iteration, hence the name Markov chain. As expected, the method is expensive and infeasible for large data sets as the complexity of the imputation increases. However, the best use of the MCMC method might be to simulate just those values needed to induce montonicity in the missing values so that the simpler regression methods might be used for imputation.

We have singled out these three methods because reliable software implementation of the techniques is available from SAS. Missing data and their treatment is an important data quality issue for data miners. A suitable solution depends on the computational resources as well as the tolerance to errors in approximating missing values.

5.2.2 Incomplete Data

There are situations where the data exist but are changed or mangled by the time they get to the user. There are two important classes of such data which are termed by statisticians as **incomplete data**.

A dataset has **truncated** data when observations are dropped from the data set. For example, customers who spend less than a dollar a year might not be included in a customer database. When such observations are dropped from a data set, they affect the sample size. Metadata and domain expertise are crucial in dealing with truncated data, otherwise it is difficult to detect that such a bias has been introduced into the data set.

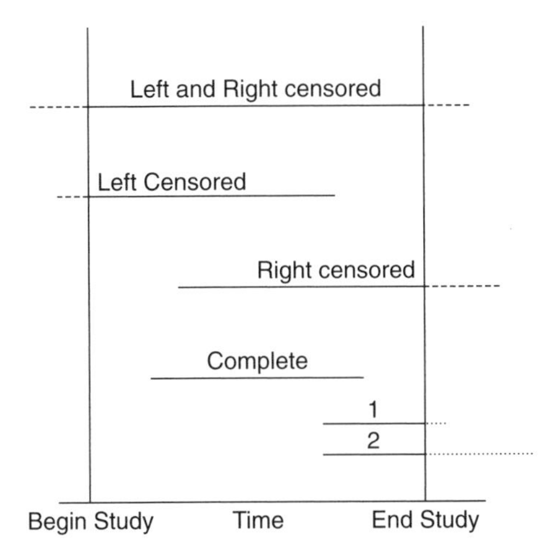

Figure 5.1: Types of censored data.

Another type of data are called **censored data**. Such data are typically studied in **duration analysis**, where times until certain types of events happen (time elapsed before the next earthquake, time until a machine runs before failing, duration before a patient's symptoms recur, time for the successful completion of a task) are of interest. Such durations have a built-in uncertainty in the sense that they cannot be predicted exactly. The focus is on understanding the probability distribution of the time intervals to be able to isolate periods of high likelihood when the event of interest (e.g., earthquake) could take place. A challenge in such data is that the intervals could be incomplete. For instance, a patient might not develop symptoms during our study and might become untrackable afterwards. So we would know that the symptoms did not occur for at least two years, but they could have recurred after two years and a week or after 10 years. We will not be able to tell the difference between the two cases, as in examples 1 and 2 in Figure 5.1. Similarly, when we measure the time to failure of a machine, we might not know when it started operating, only the time elapsed since the start of our study. The two cases we discussed are called **right censored** and **left censored** respectively. (See Figure 5.1.)

Truncation and censoring of data occur in expected as well as unexpected ways. Integer overflows might result in censored values—for example, a signed 32-bit integer can only range up to a little over 2 billion. Higher values cannot be represented. Or, the maximum duration that can be measured on a telephone switch is 24 hours. A long call is broken up into a series of shorter calls—however, we do have data to reconstruct the original data. An alarming example of censoring is when timestamps are set to defaults. In one field study, a sales order system populated missing dates on active orders with a default

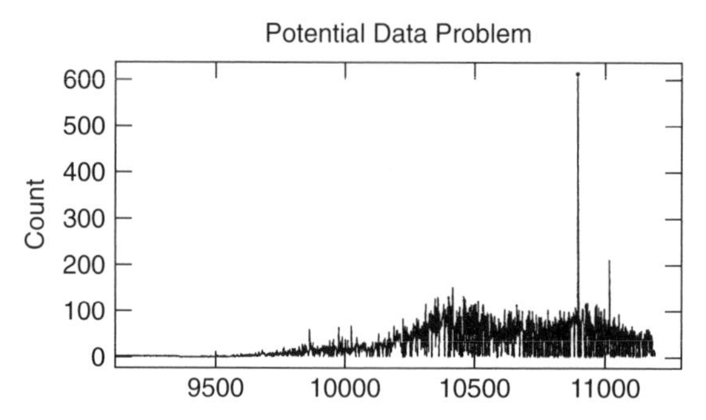

Figure 5.2: Spikes in histograms—is it censored data?

date of its choice. As a consequence, service orders with missing values all had exactly the same vintage. Such egregious instances of undocumented censoring can be detected with the help of histograms and frequency distributions. In Figure 5.2, the spike at right is a telltale sign that something is amiss, in this case, inadvertent left censoring caused by a default date. Histograms are useful because sudden changes in the distribution that appear as spikes or "V"-shaped valleys are pointers to data problems—either caused by an inexplicable preponderance of a given value (default dates, censored intervals) or data being lost or truncated.

The consequences too are unpredictable. It might be practical to drop customers who bill less than $1 per year from the data set. However, if the data analyst is unaware of the truncation or does not know the number of records truncated in this fashion, the per capita expenditure of a customer will be overstated, with eventually unpleasant consequences on Wall Street.

Given the disconnect between data collection and data mining, metadata are very important in the context of incomplete data. Otherwise, the data set is just a collection of numbers and we cannot tell the difference between an exact value and a censored value or whether the data set has been truncated. EDM techniques such as histograms and other profiling techniques are useful in detecting hidden censoring and truncation.

5.2.3 Outliers

Consider the following data:

$$3, 4, 7, 4, 8, 3, 9, 5, 7, 6, 92$$

Intuition tells us that 92 is a "suspicious" entry because our experience with the rest of the data tells us to "expect" numbers between 0 and 10. An observation that is suspicious because it is not in line with the rest of the data is an **outlier**. We will discuss precise definitions further in this section.

Outliers are important from two perspectives. They could potentially represent the consequential elements in the data. For instance, according to the informal "80–20" rule, 20% of the customer base generates 80% of the profits. Analogous rules exist in where a small percentage of root causes generate a bulk of failures in networks, software, and so on. Data miners dream of hitting this highly profitable vein of customers by sifting through the data. Such customers tend to be outliers in terms of the profits they generate. On the other hand, outliers could be data glitches. In the above list, 92 could be a simple typo where the separating "comma" is missing. Similarly, zeroes in network traffic data are almost always data glitches, especially when they occur fleetingly as in

$$10278643, 10876373, 10938746, 10298462, 0,$$
$$10289377, 10939874, 10837646, 10298274$$

or in the absence of any known outage. It is important to be able to tell the difference between outliers and data glitches. There are ways of making the distinction, as we will see further down in the chapter.

Outliers are detected by the departure of data points from what we expect them to be. The way we define "expected" and "departure" gives rise to different types of outliers. A simple approach suitable for dealing with one attribute at a time is based on error bounds. For instance, we can examine aggregates such as means of groups of data and see whether they fall within error bounds based upon standard errors and confidence intervals that we discussed in the chapter on point estimation. This approach is called the **control chart approach**. A more complex approach is to capture the interrelationships between attributes using models (e.g., regression models, generalized linear models) and detect points that are unexpected as defined by the model, giving rise to **model based outliers**. **Geometric outliers**, favored by data miners, are defined based on the geometric relationship with other points, irrespective of the density or sparseness of data. In contrast, **distributional outliers** are those that tend to be found in sparsely populated areas, not in the company of other data points. Finally, **time series outliers** require analysis based on historical data to detect unexpectedness. For example, in network data, it is common to see sudden bursts in traffic. In isolation, the bursts might seem like outliers but they might be quite consistent with the traffic patterns over time. We will discuss each of the above types of outliers below.

Control Charts

Control charts were developed in the manufacturing industry to monitor the quality of production lots. Several data samples are collected and select summary statistics (sample mean, standard error, correlation coefficient) are computed for each sample. The name of the control chart is derived from the statistic that denotes the expected value of the sample such as a **$\overline{X}$-chart** if it

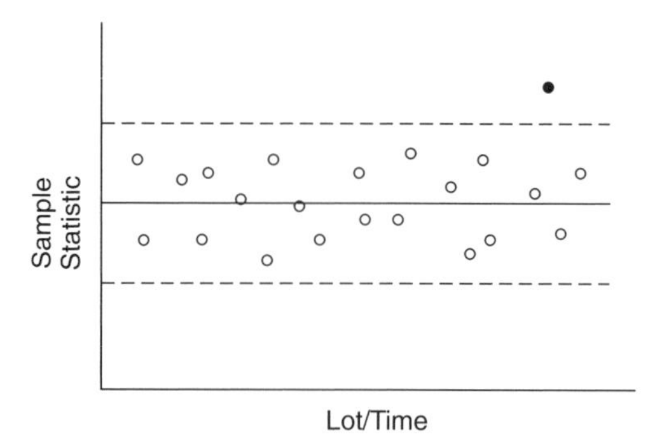

Figure 5.3: Example of a univariate control chart.

is the sample mean or the **R-chart** if the statistic computed for the sample is the correlation coefficient. The values are then plotted on a control chart, which typically consists of:

- A central horizontal line that denotes an "expected value", such as a mean or a sum (usually of binary 1/0 attributes).
- Two lines parallel to the central line that represent "acceptable bounds" for the statistic corresponding to the central line. For example, in an $\overline{X}$-chart the error bounds are based on the standard error, which represents the standard deviation of the sample mean. The intuition is based on the confidence interval discussion of point estimation which states that we can construct intervals within which the sample mean will lie for approximately 95% of the samples. The bounds are also called **sigma-limits** since they are based on the sample standard deviation, commonly denoted by σ.
- The statistic corresponding to each sample is plotted, with the sample ID (historically corresponding to a manufacturing lot number) on the X-axis and the sample statistic on the Y-axis.

Any sample whose statistic falls outside the error bounds is said to be **out of control**. Figure 5.3 shows an example where the X-axis corresponds to the time when the sample is drawn. The plot shows one data point in the upper right-hand side that is outside the bounds. This particular data point is an outlier, a good candidate for further investigation.

Control charts are easy to compute, understand, and explain. Since they are based on aggregates, they scale well to large data sets. But notice that the computation of the error bounds depends on the sampling distribution of the statistic used to capture the expected values. The mean is ideal since its sampling distribution does not depend on the distribution of the underlying data. (See Chapter 2.) However, other statistics such as the correlation coefficient require

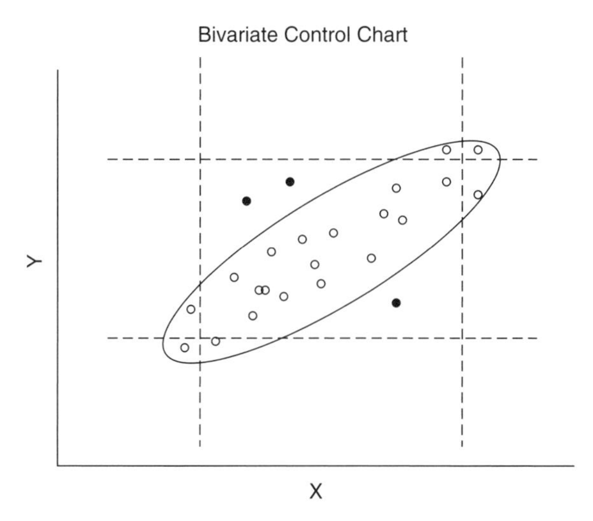

Figure 5.4: Example of a bivariate control chart.

assumptions about the underlying distribution of the data (e.g., multivariate normal for the R-chart), which is not in keeping with the nonparametric nature of data mining.

Control charts are primarily suitable for studying one or two attributes at a time. They cannot be used for capturing outliers based on interrelationships between attributes. It is quite possible that a data point might be well behaved with respect to any given attribute, but out of control with respect to the attributes taken together. In Figure 5.4, the region inside the rectangle represents attribute-wise control limits whereas the region inside the ellipse represents the joint control limits based on taking into account the joint distribution of the two attributes. Clearly, the two regions are not identical. Therefore, a point could fall outside the real joint control region and be a potential outlier, yet could fall inside the rectangular region and be interpreted as "in control" on an attribute-by-attribute basis.

There has been some work on bivariate control charts like the one in Figure 5.4. While there is not much work on multivariate control charts, there are two interesting approaches. The first approach by Liu and Singh maps n-dimensional data to a single dimension *depth* and constructs univariate control charts based on the attribute depth. The second approach is derived from the field of chemical process control and uses wavelet representations to detect outliers that occur at multiple scales.

Model Based Outlier Detection

Interrelationships between attributes are captured using parametric models such as linear regression, logistic regression, and others. Such representations are simple and can be computed from a small amount of data. (The

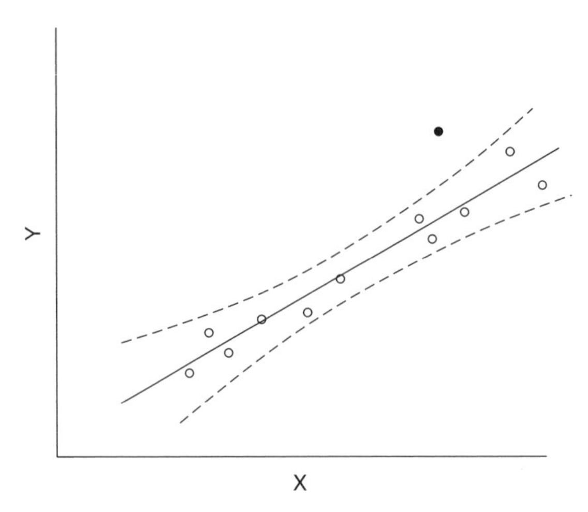

Figure 5.5: Finding outliers using linear regression.

appropriateness, however, depends on whether or not the distribution assumptions are correct.) If a model represents a large portion of the data and captures general attribute relationships, we can identify data points that do not conform to the models. Linear regression provides a simple example.

In Figure 5.5, the solid line represents the linear regression model that represents the majority of the data. The dotted lines represent the confidence bounds. We can see that there is one point that lies outside the confidence bounds dictated by the model, a potential outlier. There are well-known methods of detecting outliers with respect to regression models, and which are available in standard software such as SAS. Such methods compute the difference between expected values (as per the model) and observed values and tag data points where the difference is outside that of statistical error. The differences, called **residuals**, can be analyzed further to reveal any systematic biases and patterns (e.g., **heteroscedasticity**—the variance of an attribute depends on the values of another attribute) that exist in the outliers.

Set comparison is a major application of model-based outlier detection. We discussed this in Section 3.5. Set comparison techniques can rapidly isolate sections where the data differ, enabling us to use more sophisticated techniques to analyze just the suspect or "outlying" sections. The set comparison technique can be used to compare a test data set to a baseline data set that has been certified to be clean, to identify sections in the data that are different from the clean data set.

Another technique is to use piecewise linear regression (or any piecewise summarization of the data using parametric and nonparametric models) by fitting linear regression models within each partition and using the model parameters to compare and detect abnormal sections of the data. We discuss below an example of detecting glitches through set comparison.

5.2.4 Detecting Glitches Using Set Comparison

We conducted two experiments to illustrate glitch detection by knowingly corrupting a data set. The first experiment simulates the switching of two fields that are similar in their marginal behavior but differ in their interactions with other variables. The second experiment mimics two errors, incomplete processing of a variable and the censoring of another.

We generated a clean data set D_g of 62,500 observations, each with three variables Wait, Length, and Fee. All three are numeric variables. The first two are negatively correlated but have very similar marginal distributions. They are uncorrelated with the third variable Fee. The baseline DataSphere parameters which serve as a benchmark for comparison are computed using this data set.

Switched Fields

For the first experiment, we created a corrupted data set (denoted D_t) by switching fields Wait and Length in our original data set (denoted by D_g). We then applied the set comparison analysis of Section 3.5, please refer back to that section for details.

There are 43 populated classes of the DataSphere partition of D_t. The chi-square for the Multinomial test is 56.16. The p-value for the χ^2 at $42 = 43 - 1$ degrees of freedom is 0.07. That is, the χ^2 from the Multinomial test shows significant differences at 90% level of confidence. This test indicates a possible difference between D_t and D_g in the distribution of points among the partition classes, so we performed the Mahalanobis test for difference in multivariate means of D_t and D_g each DataSphere partition class. There were 43 such tests, one for each populated DataSphere class.

The Mahalanobis test identifies many classes where the multivariate means of the corresponding DataSphere classes are significantly different; see the bubble plot in Figure 5.6. The X-axis represents the distance layers where negative layers correspond to negative pyramids. The Y-axis represents the pyramid. For example, the tuple (–5, WAIT) represents the class in the DataSphere partition corresponding to distance layer number 5 (from the center) and the pyramid of the attribute WAIT, where the deviation from the center is maximum for the attribute WAIT and the deviation is below average (negative sign) with respect to WAIT. The presence of a bead indicates a significant difference in the multivariate mean of the DataSphere class (X, Y) between the data sets D_g and D_t. There are many differences all across the dataset as a result of switching the fields.

Improper Processing and Censoring of Variables

In the second experiment, we introduced two glitches. First, a volume charge that was applied to obtain Fee in the good data set D_g (only very large values of Wait qualify for the penalty) was not applied for D_t. Second, measurements of Length were censored so that any value below 0.05 was set to 0.05 and any

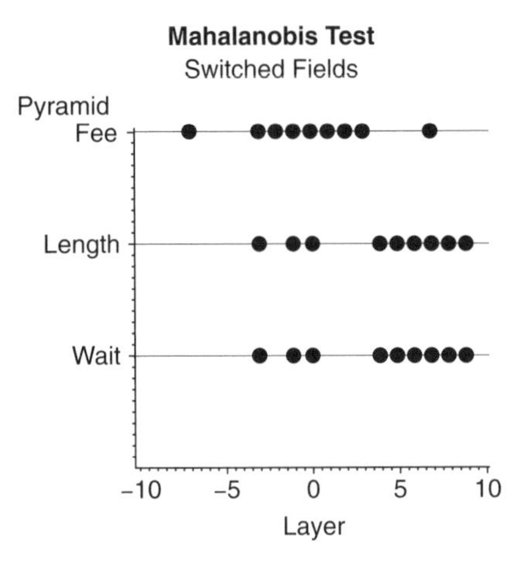

Figure 5.6: Comparison of D_g and D_t.

value above 20 was set to 20. In short, the glitches affected a small proportion of data in the tails of the distribution.

The multinomial test came out to be not significant since the corrupted points fall in outlying layers in both the good and bad data sets. However, the Mahalanobis test of DataSphere section centers identifies the problem segments. The multivariate means for the corrupted sections of the data are significantly different and are immediately identifiable on the bead plot shown in Figure 5.7.

We note that univariate tests do not identify any differences in any of the variables, in any of the sections, even after partitioning the data using DataSpheres.

Once the corrupted classes have been identified, the individual data points that fall into those classes can be extracted and studied further to determine the cause of the difference. Figure 5.8 shows the univariate box plots of data points in the subset identified by the Mahalanobis test. It is clear that the variable Wait is unchanged, while Length is censored at the tails and the distribution of Fee is less spread out for the "BAD" data set when compared to the "OK" data set. Since we could isolate the corrupt classes, we could examine them closely with methods suitable for small data sets.

Geometric Outliers

Data points that are on the periphery of the data set can be termed **geometric outliers**. A natural intuition for geometric outliers can be seen in Tukey's definition of **peeling**. We can think of the data set as consisting of layers (like an onion) and we peel off layer after layer to go from outlying data to the very heart of the deepest reaches of the data set. There are two approaches to peeling: (a) we can peel off successive **convex hulls** (i.e., a polygon of minimum

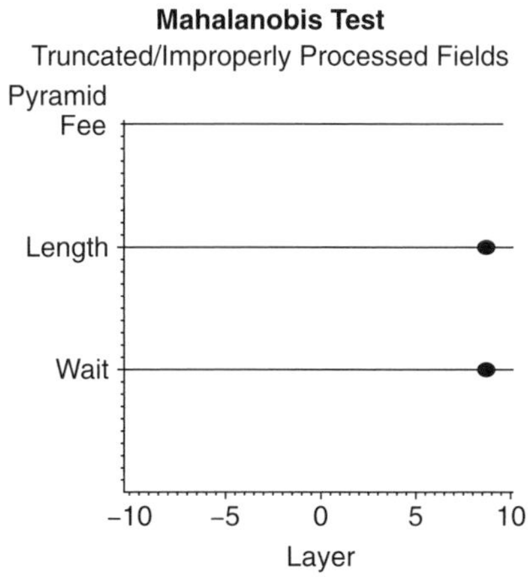

Mahalanobis Test

Truncated/Improperly Processed Fields

Pyramid Fee

Length

Wait

−10 −5 0 5 10

Layer

Figure 5.7: Comparison of D_g and D_t.

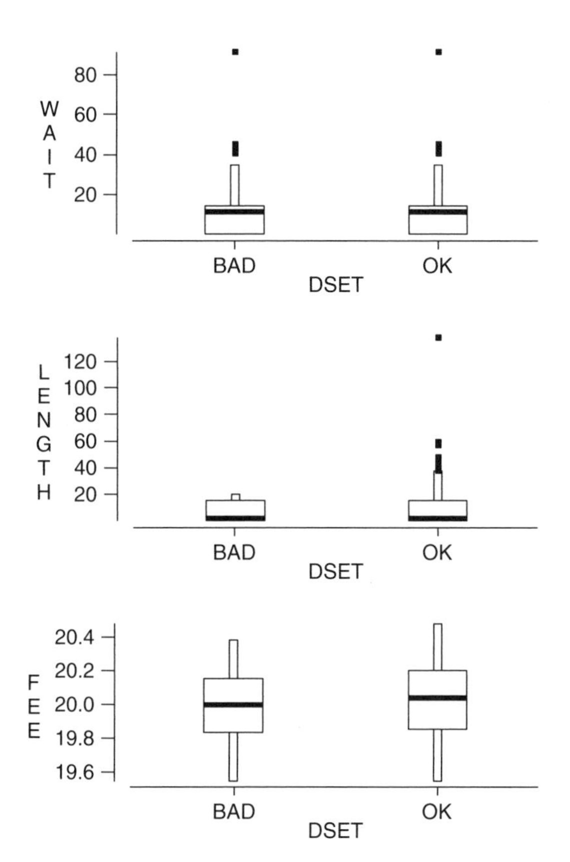

Figure 5.8: Comparison of data in segments that are different.

area that contains all of the points) where the external convex hulls would ideally contain the outliers. (b) we can compute depth contours, perhaps in a center outward fashion and define depth contours as data points with the same depth. The depth contours with the least depth would potentially contain the outliers.

As expected, the computation of convex hulls and depth contours is expensive and cannot be done in more than three dimensions as of now. While there are efficient algorithms for 2-D and 3-D the lower bound on the computational complexity increases rapidly at the rate of

$$\Omega(N^{ceil(d/2)}),$$

where N is the number of data points and d is the number of dimensions.

Distributional Outliers

Distributional outliers are defined as those points which are in a region of "low density". Since these points are relatively isolated, there is a good chance that they are outliers. One way to define distributional outliers is to compute, for every point x in the data set, the value

$$DB[D](x),$$

which is the fraction of points in the data set at distance D or greater from x. Then, the set of $DB(p, D)$**-outliers** is the set of points $\{x\}$ such that $DB[D](x) > p$.

The intuition behind this definition is that outliers are likely to be at a large distance (appropriately defined, e.g., L_2) from the other points. However, the outliers might be clustered, perhaps because of default or censored values for some of the fields. Therefore, p can be adjusted to account for these clusters.

An advantage of this definition of distributional outlier is that is fairly nonparametric, although some tweaking of p, D, and the distance function is required. Another advantage is the availability of fast algorithms for computing the $DB(p, D)$ outliers. One method is to compute the distance from every point to every other, resulting in a $O(dN^2)$ algorithm, where d is the number of dimensions and N is the number of points. Another method is to partition the space into c^d segments, such that every pair of points in the same cell is within distance D of each other. Then, points in neighboring cells are compared to each other, resulting in an $O(c^dN)$ algorithm.

5.2.5 Time Series Outliers: A Case Study

Longitudinal data, or time series data, must be analyzed in a different manner from snapshot or cross-sectional data. First of all, data that are close together in time tend to be highly correlated. We are often interested in identifying these correlations. Furthermore, there could be cycles in the data, for example, long distance telephone usage has weekday morning and afternoon peaks. In order to determine whether a data point is an outlier or not, it is important to check its past behavior. We will discuss this point in some detail using a case study because we feel that there is no strong reference for the discussion of this material.

We discuss a two-step approach to detecting outliers in time series that could be potential data glitches. We build the glitch detection mechanism using the EDM summaries that we discussed in Chapter 2. First, we divide the attribute space into sections using a space partitioning strategy, in this instance, the DS technique. We treat each class of this partition as a *state* that a data point can be at any point in time. A given time series can then be expressed as a trajectory of the states. The trajectories can be characterized using transition probabilities that are estimated from the data. At any point in time, transitions can be ranked by their likelihood. "Low-likelihood" transitions are outliers that should be flagged as **data alerts**. The data alerts can be further analyzed to separate abnormal but legitimate behavior (bursty traffic) from data glitches (missing data). Since the data alerts constitute a small subset of the original data, statistical methods intended for smaller data sets could be used to separate the truly bad data. In the second step, we map the multivariate time series to a time series of two deviation attributes. We then define conditions under which the deviations are flagged as abnormal.

The first measure of deviation is the **Relative Deviation**, which represents the movement of a data point relative to other data points over time. For example, an online customer might be purchasing merchandise at a faster rate than others. Another customer might continue at the same rate at which he or she started. The trajectories of purchases of these customers will be different.

The second measure of deviation is the **Within Deviation** that measures whether a data point is different at any given time t with respect to its own *expected behavior*. The latter can be defined in several ways, depending on the resource constraints. A simple strategy would be to fit a linear model to the time series of a given record using summaries and identify departures from the model.

Note that the relative deviation is more robust, since it is difficult to change state (i.e., position in the attribute space relative to others) without a significant change in the attributes. The relative deviation serves an additional purpose of identifying the data point as typical (states that are in the inner distance layers) or atypical (in the outer layers). In contrast, the within deviation is very sensitive to minor changes and is better for capturing long-term trends of the individual data point. Due to this property, we can use the within deviation to differentiate between legitimate changes and data glitches as discussed later in this section.

Example—ATM/Frame Relay Data

We used a data set that measured four attributes for a type of data service "connection", namely, Bytes Received, Bytes Transmitted, Frames Received, and Frames Transmitted, over a 31 day period. There were 15,596 connections that were observed daily. The data consisted of the daily totals of the four attributes during the 31-day period. We computed the within deviation of a point at time t simply to be the sum of the standardized deviations of the individual attributes

$$dev_i(t) = \sum_{j=1}^{d} \left(\frac{\left(x_{ij}(t) - \overline{x_{ij}}\right)}{s_{ij}} \right)^2,$$

where $x_{ij}(t)$ is the value of the jth attribute of the ith data point (in this case the connection) at time t, $\overline{x_{ij}}$ is the 31-day average of the jth attribute for the ith data point and s_{ij} is the standard deviation of the jth attribute for the ith individual over the 31-day period.

We used the connection average for 31 days to create a data partition using DataSpheres with 4 layers and 8 pyramids, 2 pyramids for every attribute. For the purpose of simplification, we collapsed all the negative pyramids into a single "negative" orthant and all the positive pyramids into a single "positive" orthant within each of the 4 layers.

Next, we computed the transition matrices using the sample proportion

$$\hat{P}(i,j,t) = \frac{n_{ij}(t)}{n_i(t)},$$

where $P(i, j, t)$ is the probability of changing from state i to state j at time t ($\hat{P}$ denotes an estimate), $n_i(t)$ is the number of points in state i at time t, and $n_{ij}(t)$ is the number of points that move from state i at time t to state j at time $t + 1$.

We noticed that the estimated probability of changing states at any point in time was less than 0.25, usually less than 0.2. Therefore, we extracted the entire multivariate time series for every data point that changed its state at least once in the 31-day period. That is, we used *"change in state" to flag data alerts*. The change in state happens when a data point crosses a class boundary which, in turn, is a function of all four attributes. Note that if data points changed states more frequently we could have defined the most likely transition state(s) whose transition probabilities add up to some threshold α (say 0.80) and flag all other state transitions as abnormal (yet another way to define the relative deviation).

We also noticed that there was one distinctive feature that set the data problems apart from the changes caused by abnormal but genuine events, namely a successive flip flop in states. For example, the transition sequence of states $i \rightarrow j \rightarrow i$ over three consecutive time steps. This sequence was usually caused by missing data or a short-term outage that caused the traffic to drop.

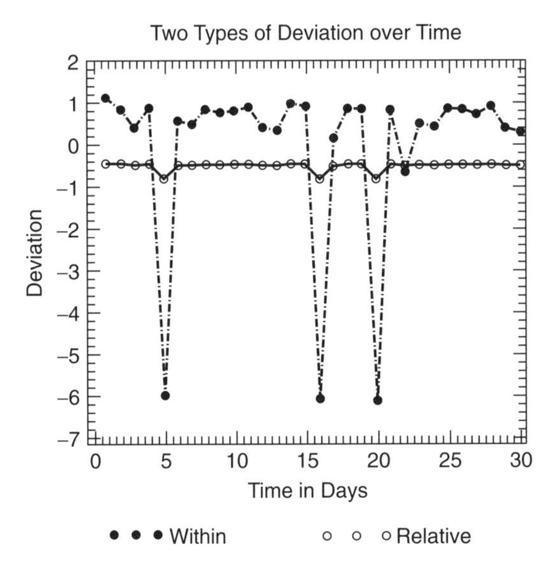

Figure 5.9: Data problem indicated by flip-flop of states.

To illustrate, we plotted four representative types of "abnormal patterns". In Figure 5.9, the data problem is indicated by a consecutive flip-flopping of states indicated by the relative deviation. (For the purpose of plotting alone, we have used a suitable transformation of the change in state variable to denote relative deviation.) Note that the within deviation is much more volatile. The fact that the within deviation drops to the same level on each of the three times large variations indicates that the values are being set to some default (such as zero) due to missing data.

In Figure 5.10, there is significant volatility indicating occasional bursts of activity. However, note that the within deviation at time 21 is in the opposite direction, dropping to a default value indicative of missing data.

In Figure 5.11, the behavior is quite different. The drop in the two deviations is indicative of migration of usage to other services or carriers, with occasional dribbles of traffic. Note that the flip-flop at times 15 and 30 correspond to within deviations of different amounts indicating that they are genuine bursts of traffic rather than data problems.

Finally, Figure 5.12 seems to indicate a genuinely volatile customer. The flip-flop at time 17 is probably legitimate and not a data problem. This could happen when a data point is close to the partition class boundary.

We have used the within and relative deviations to isolate a handful (20%) of the data as potentially "dirty". We note that automatic detection techniques such as logistic regression or machine learning or clustering are expensive and

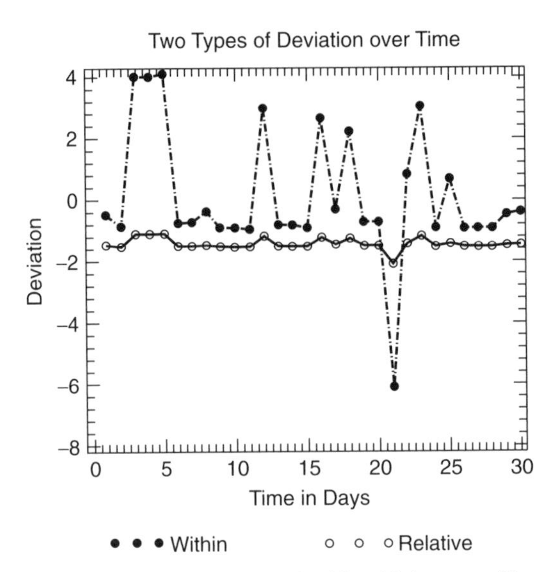

Figure 5.10: Data problem mixed in with bursty traffic.

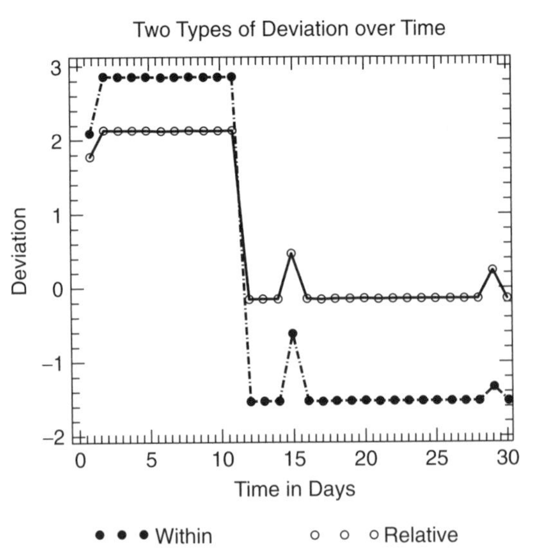

Figure 5.11: Big shifts in behavior.

will not be effective on noisy data. Additionally, note that we have built up this effective mechanism of glitch detection in massive time series data using the simple EDM summaries like mean, proportion and standard deviation that we discussed in Chapter 2.

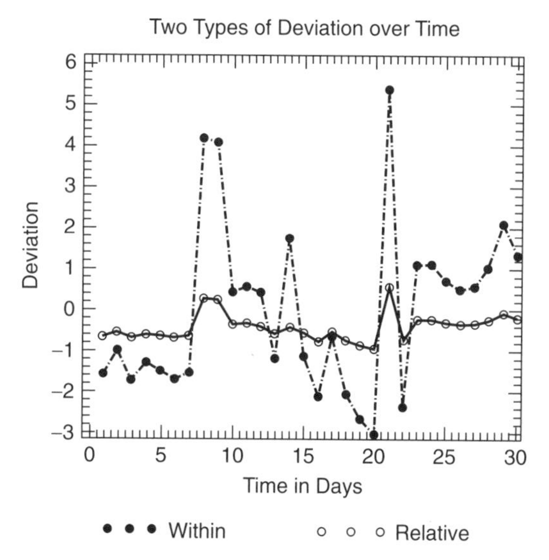

Figure 5.12: Legitimate volatile behavior.

Glitches or Legitimate?

Once the outliers are identified, there are two major issues we need to address. First, we need to distinguish between real data problems and genuine but atypical changes in the data. Second, we need to define the action to be taken with respect to the data glitches.

How can we tell the difference between an outlier and a legitimate change in the data? We propose below broad guidelines:

- Genuine changes are usually persistent over time, whereas data problems appear and disappear quickly.
- Arbitrary data glitches tend to appear randomly without any structure while glitches with a reason can be "rationalized". Even though they can be rationalized, they are still glitches because they are unintended. For example, a geographical proximity in the glitches would suggest a systemic cause such as a drought resulting in lower crop yields in that region. Similarly, a drop in revenues at a single point in time is more likely to be a data problem (missing data) than a sustained downward trend. Note that patterns in glitches can reveal systemic causes such as data from a particular segment being missing.
- We can use the within deviation of a data point to separate out differences with "structure" (systemic changes in the process that generates the data, resulting in shifts in the distribution) as opposed to random aberrations. Using the departure from linear autoregressive models as measures of within deviation is a potential way of detecting structure.

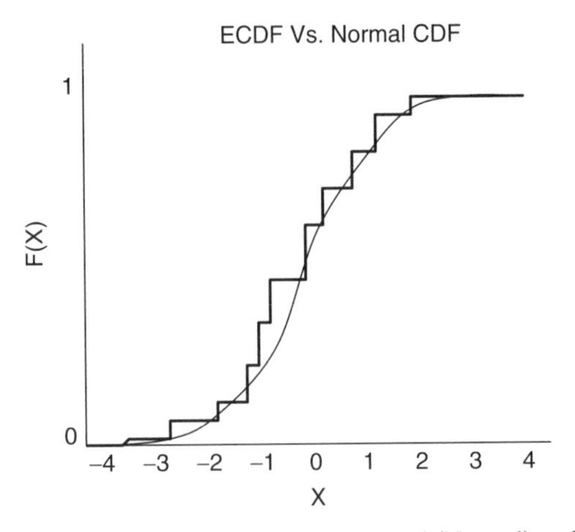

Figure 5.13: Measuring the distance between an assumed (Normal) and empirical CDF.

We can use the methods discussed in process based data quality techniques and missing value imputation to deal with glitches and outliers detected in time series data. They can be dropped, studied separately, or substituted with more reasonable imputed values.

5.2.6 Goodness-of-Fit

Statisticians have used "goodness-of-fit" tests to check whether there is a good match between the data and the model or the hypothesis used to represent the data. Prominent among these are the chi-square test for testing the hypothesis of independence of attributes and the Kolmogorov-Smirnov test for testing how well a specified distribution represents the data based on the empirical cumulative distribution function that we discussed in Chapter 3.

In the context of data quality, we can use goodness of fit tests to validate the analysis techniques. For example, a common starting point for many techniques is to assume that the data have a specific distribution say Binomial or Normal. Everything that follows hinges on the assumption being correct. Is such an analytic assumption suitable for the data? For a discrete distribution (such as the Binomial) we can use the chi-square test. Suppose we expect that the probabilities in the d categories should be $p_1, p_2, \ldots, p_d$, as per the assumption, then we would expect that in category i, there should be np_i data points. However, our data indicate that there are n_i data points. Then

$$\sum_{i=1}^{d} \frac{(Observed - Expected)^2}{Expected} = \sum_{i=1}^{d} \frac{(n_i - np_i)^2}{np_i}$$

has chi-square distribution with $d - 1$ degrees of freedom. We discussed this in Section 2.4.4 in the context of equation 2.31. We can establish whether this is a reasonable value by comparing the above value with the value from an appropriate chi-square distribution. If the values are too far apart, then our assumption is incorrect. The specified values $p_1, \ldots, p_d$ are obtained either from past experience or by assuming a distribution like the Binomial and computing the probabilities. (Note that these are univariate tests, though we discussed a multivariate version based on the DataSphere bins in Chapter 3.)

For continuous distributions such as the Normal distribution, the Kolmogorov test measures the maximum distance between the assumed distribution and the empirical distribution computed from the data. More complex goodness-of-fit tests exist for models such as logistic regression, generalized linear models, and others. The $Q - Q$ plots that we discussed in Chapter 3 are another good diagnostic tool for verifying goodness of fit between assumptions and data. Many software packages offer goodness-of-fit statistics. SAS has extensive coverage in its PROCs UNIVARIATE, NLIN, LOGISTIC, GENMOD and others.

We can define the goodness-of-fit of a given regression model, as measured by regression depth (see Section 2.9.1). Intuitively, a regression line (plane) has greater regression depth if it has to pass through many data points to rotate to the completely vertical position which represents a "non-fit", since it implies no relationship between the response variable and the covariate(s). The greater the regression depth of the model, the better it is at representing the data.

Choosing a model that reflects the true nature of the data is an integral part of the data quality process. The quality of the results from a data mining exercise depends both on the reliability of the data as well as the appropriateness of the assumptions and models used in the analysis. Therefore, it is important to spend time using goodness-of-fit tests to verify the suitability of the assumptions and models used to analyze the data.

5.2.7 Annotated Bibliography

Designing experiments for collecting data is described in R. A. Fisher's book [46]. Please see [81] for a discussion of statistical analysis with missing data including imputation through regression. In particular, see [108] for a discussion of the role of the propensity scores. Please see [113] for a discussion of the MCMC method of imputation. Please see a SAS white paper [129] for a summary of the missing value imputation techniques and explanation of the use of SAS procedures MI and MIANALYZE mentioned in [66]. An interesting way of discovering glitches in massive time series data using a Markov chain representation is discussed in [32]. In this paper, missing values in time series are isolated by searching for *flip-flop* patterns. Other references for dealing with missing data include a tutorial presented at the SIAM International Conference on Data Mining in 2002 [100].

Failure time data and their analysis is covered in [74]. Please see [75] for an overview of failure analysis techniques for censored and truncated data.

An introduction to quality control in industry and the use of control charts and statistics in given in [40]. A bivariate approach to control charts is discussed in [1]. A depth-based approach to multivariate process control is discussed in [83]. Discovering departures from expected bounds for errors occurring at different scales is discussed in [3].

Departures from expected values as determined by models can be identified by analyzing the residuals (the differences between observed and predicted values). Please see [5] for a discussion of regression diagnostics. Please see [111] for the computation of regression depth of points in high dimensions.

The concept of "peeling" off layers of data is discussed by Tukey in [123]. Peeling by convex hulls is discussed in the book by Preparata and Shamos [101], which is an excellent introduction to geometric methods. Johnson, Kwok and Ng [73] and Miller et al. [88] give fast algorithms for computing depth contours.

The definition of $DB(p, D)$-outliers and fast algorithms for computing them are found in [77]. This idea is extended to *local* outliers in [12].

Goodness-of-fit tests such as the chi-square and Kolmogorov-Smirnov for changes in distribution are discussed in [27]. More complex tests for regression models and generalized linear models are discussed in [86] and [105].

5.3 DATABASE TECHNIQUES FOR DQ

A large fraction of today's data is stored in modern relational database management systems (DBMS). This situation is fortunate because a modern relational DBMS has extensive facilities for ensuring data quality and documenting metadata.

5.3.1 What is a Relational Database?

A **relational** database is a collection of **tables** (or **relations**). Each table is an unordered collection of **records**, and each record has a collection of named fields. In turn, tables are collected into **tablespaces**. Figure 5.14 shows an example database layout. At the top level is a collection of tablespaces (Sales, Provisioning, etc.), each containing a related collection of tables. For example, the Sales tablespace contains a number of tables, including Salesforce and Orders. The Salesforce tablespace contains fields which describe individual members of the sales force, including the name of the salesperson, and their base salary and sales commission rate.

A relational database thus consists of a collection of tables. One of the key ideas in building and using relational databases is to **join** tables together to merge the information stored in different tables. For example, one of the fields in the Orders tables is the SalesforceID (denoted by Orders.SalesforceID),

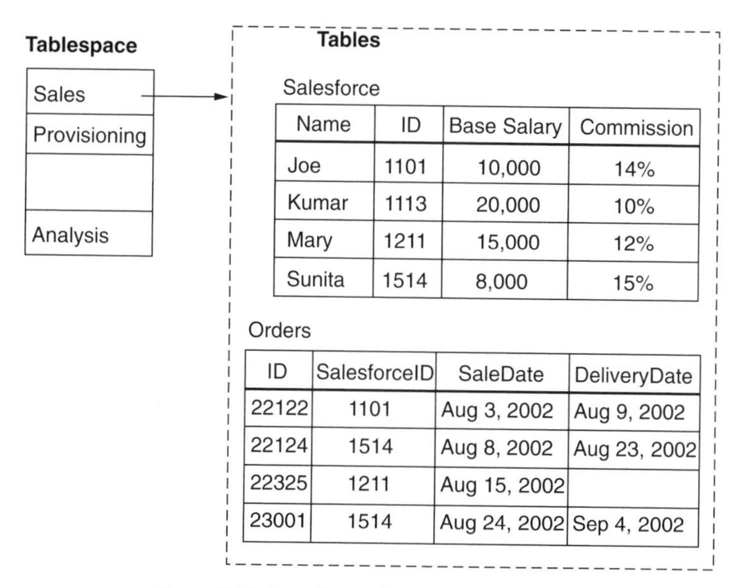

Name	ID	Base Salary	Commission
Joe	1101	10,000	14%
Kumar	1113	20,000	10%
Mary	1211	15,000	12%
Sunita	1514	8,000	15%

ID	SalesforceID	SaleDate	DeliveryDate
22122	1101	Aug 3, 2002	Aug 9, 2002
22124	1514	Aug 8, 2002	Aug 23, 2002
22325	1211	Aug 15, 2002	
23001	1514	Aug 24, 2002	Sep 4, 2002

Figure 5.14: Data layout in a relational DBMS.

which indicates which member of the sales force negotiated the sale. The field Salesforce.ID is a **primary key** of the Salesforce table, meaning that it uniquely identifies the member of the sales force. (The field Name is also unique in every record of this instance of Salesforce, but there is no guarantee of uniqueness, for example, if another salesperson named Joe was hired.) Because the field Orders.SalesforceID refers to the primary key of Salesforce, it is a **foreign key**, and the association of Orders records to Salesforce records through the association Orders.Salesforceid = Salesforce.ID is a **foreign key join**. That is, Orders.Salesforceid and Salesforce.Id are **match keys** or **join keys**. A database in which redundant information (such as the salary of the salesperson who negotiated an order) is factored out to separate tables is said to be **normalized**.

One important way in which a relational DBMS can help to ensure data integrity is through the design of the database. We can relate a salesperson to all of their sales through the foreign key join—we do not need to record this information on a per-order basis. By recording the information about a salesperson only once, its consistency throughout the database is ensured, and we are better able to keep its values correct.

Another important facility for ensuring data quality is the use of **transactions**—which ensure that database updates (and queries) occur in an isolated and all-or-nothing fashion. For example, each order might consist of a number of items, stored in the table OrderItem and related to the Orders table by the foreign key join OrderItem.OrderID=Orders.ID. Suppose that Kumar makes a sale for 5 Snarks and 4 Gryphons. Recording this order requires three database updates, one apiece in OrderItems for the Snarks and the Gryphons, and one in Orders for the order itself. If any of these three updates is not performed, the database becomes inconsistent.

A modern DBMS can also enforce a number of data integrity **constraints**:

- Every field in a table has a **data type** as well as a name. Common data types are string (text), integer, floating point, and date. An attempt to store "Bob" in a date field would be rejected as an error.
- The values of a field can be further restricted. For example, a field which stores a person's Age could be an integer restricted to the range 0 through 150. String values are often constrained to satisfy a regular expression. For example, a constraint on TelephoneNumber might be that it satisfy a regular expression such as "([0–9]{3}) [0–9]{3}–[0–9]{4}" (three digits, a space, three digits, a dash, four digits).
- We might often work with special data types, such as telephone numbers, social security numbers, and so on, and use them in many fields. To ensure a consistent definition of the field constraint, we can define a **domain**—essentially a named field constraint.
- If we do not know the value of a field, we can specify that it is NULL (e.g., order 22325 in the Orders table of Figure 5.14 has a NULL DeliveryDate). We can specify whether entries of a field can be NULL or not. Similarly, we can specify which fields or field combinations are keys (e.g., they must be unique in every record).
- We can specify the foreign key-primary key associations, and the actions to be taken in case of a violation (ranging from issuing a warning to cascading deletions).
- We can specify application-specific constraints through the use of **triggers**, which specify an action to occur whenever a constraint is satisfied. For example, a trigger can send an e-mail to the head of sales whenever a salesperson negotiates a sale for a price lower than production costs.

A relational database will have a data querying and manipulation language which is a variant of standard SQL (Structured Query Language). SQL is a powerful query language which can express many complex reporting tasks in just a few lines. Many DBMSs have further reporting tools which make data analysis even easier, for instance, a point and click graphical interface.

A database is part of a larger system in which users interact with the database in a variety of ways to accomplish their tasks. Many of these tasks can be written in SQL and, in fact, stored in the DBMS and made available to authorized users:

- The database can contain **stored procedures**, which encode a combination of SQL and procedural logic to accomplish specialized tasks. One stored procedure can be a simple data transformation (e.g., to change the format of a telephone number from (123) 456-7890 to 1234567890), or a complex procedure involving multiple database queries and updates (e.g., to check if an order can be provisioned, and if so, reserve the resources).

- The database can store queries (called **views**) over its tables, and make them transparently appear as tables to the users. For example, a view might associate the related Salesforce record with each Orders record, giving the head of sales a more convenient view of the sale force's performance.

Finally, a modern DBMS will store a large amount of **metadata**, including:

- For every field, the field name, its data type or domain, any additional constraints, and comments about the meaning of the field.
- For every table, the table name, its keys and primary keys, any additional constraints, and comments about the meaning of the table.
- For every tablespace, comments about the meaning of the tablespace.
- The set of all primary key to foreign key associations, and other inter-table constraints such as triggers.
- The set of all stored procedures and views.
- An **audit log** of all transactions issued by the users (with an association to the user who issued the transaction).
- Other information such as character sets, user authorization lists, and so forth.

5.3.2 Why Are Data Dirty?

Given the many data integrity and metadata resources provided by a DBMS, it would seem that all data quality problems should have been resolved by now. Why is it that all the data we have analyzed, including much accessed directly from the DBMS, have significant data quality problems?

There are as many reasons as data sets, but some common reasons are:

- A significant quantity of data is stored in legacy (pre-relational) databases, flat files, spreadsheets, log files, and ad-hoc structures.
- Metadata and data integrity are lost when data is transferred from the database, for instance, in flat files representing the output of a view.
- Incorrect data are entered, for example, typos, and so on.
- The DataBase Administrator (DBA) does not understand how to apply the DBMS data integrity features, or otherwise does not use them. One problem is that setting up the features can be time consuming and unrewarded (e.g., adding comments about the meaning of each field). Constraint checking can be very expensive (e.g., foreign key join constraints and triggers), causing the DBA to turn them off. Similarly, it can be easier

to disable constraint checking than to diagnose the cause of data quality problems.

- The database can be so complex that the DBA cannot figure out the proper set of constraints to use, or the proper database structure. For example, the DeliveryDate field in the Orders table might be NULL until the order is fulfilled. What happens if a decision is made to allow partial fulfillment of orders? The meanings of many fields, tables, views, and stored procedures must now be changed. Some constraints can have unintended consequences, such as preventing the entry of valid data (e.g., a database might expect that everyone has a first and last name, but some people in the world have only one name, or more than two names). Enforcing the foreign key join constraint can cause cascading deletes. For example, suppose that Joe is fired, and his record is deleted from the Salesforce table. To ensure the foreign key constraint on Orders.SalesforceID, all of Joe's sales are deleted from the Orders table—not what was intended.
- Many business processes involve interactions among multiple organizations (e.g., sales, warehouse, billing), and therefore multiple databases. The interaction between these organizations and their databases is very frequently a cause of problems. One issue is that few DBMSs support transactions between different DBMSs, so many interactions are recorded at one database but not another. (this problem can be mitigated by using "persistent queues" to store pending updates.) Another problem is a lack of communication between the organizations, so that consistency constraints across the multiple databases is not maintained.
- Sometimes problems are caused by data cleanup efforts! For example, a duplicate elimination effort (see Section 5.3.4) might decide that two different customers are the same—and thereby "lose" one of them. In general, acting on incomplete knowledge is a dangerous thing.

Many of these problems are not database specific, and other chapters of this book address problem sources and solution techniques. In this section, we discuss some database-specific tools, namely, ETL tools, database profiling, and approximate join techniques.

5.3.3 Extraction, Transformation, and Loading (ETL)

Extraction, Transformation, and Loading (ETL) refers to the process of loading data into a database, usually from a less-than-reliable source. The data might be difficult to extract, and once extracted, in the wrong format. As can be imagined, data obtained via an ETL tool are likely to be quite problematic. However, a wide variety of research has been performed and many companies sell software to help mitigate these problems.

Extraction

Extraction tools make accessing data from a variety of sources easier. An extraction tool might be as simple as a driver for reading a delimited ASCII or ODBC data source. More complex tools can extract data from legacy data sources, for example, IMS databases. Other complex tools submit queries and perform screen scraping (e.g., simulate a 3270 terminal to extract data) or HTML scraping.

The data sources are often provided in a structure which is incompatible with the desired output. The data source might provide a limited query interface, or provide a set of tables which must be joined or otherwise transformed to provide the desired answer. In this case **mediators** and **schema mapping** tools can help to find a strategy for extracting the desired data.

Given the vast amounts of information published on the Web, it is a very attractive data source. However scraping web data is difficult for many reasons—HTML is a text markup language rather than a data publishing language, the layouts change rapidly, and many data sources try to make scraping difficult. Much research and many software companies have produced tools to help automate web scraping.

A common problem with scraped data is the lack of a **schema**, or data layout description. Some tools have facilities which can guess the meaning of a field. The main idea is to define a collection of domains (e.g., name, address, telephone number, quantity, etc.) and a set of regular expressions for each of the domains. For example, a set of regular expressions for a telephone number could include { [1]?[0–9]{10}, ([0–9]{3}) [0–9]{3}–[0–9]{4}, . . .}. Each value of a field is matched against the regular expressions for each of the domains. When the field value matches one of the regular expressions, it "votes" for the domain (a field value can vote for a domains only once, but it might vote for multiple domains). The domain with the highest vote count wins, but of course, tight elections require a close examination of the dimpled chads. Obviously, such a tool is useful in many settings.

Transformation

After extraction, the data are probably not in the desired format. An ETL tool typically provides a collection of transformation services. Many of these are simple field transformation, for example, extracting a first name and last name from an input name field, normalizing telephone numbers to the format (123) 456-7890, and so on. Other transformations are more complex, and can involve join queries, or even more complex transformations such as pivots (which turn fields into sets of records and vice versa).

Loading

ETL tools facilitate database loading by certifying that the records to be loaded already satisfy the DBMS's integrity constraints. This allows the DBA to use the "fast load" option in the database loader, which can speed database loading by an order of magnitude. The ETL loader can also use fast bulk algorithms for testing more complex constraints, such as foreign key constraints. In general, a good loading component of an ETL tool can eliminate performance concerns for enforcing data integrity constraints.

5.3.4 Approximate Matching

In Figure 5.14, the method for joining Orders with Salesforce is an precise join using the predicate Orders.SalesforceID = Salesforce.ID. We are assured that every value of Salesforce.ID is unique and that it has been accurately transcribed in Orders.SalesforceID. This case is taken to be the norm by most database literature.

However, we are often faced with the need to associate two tables without the benefit of a precise join key. One example occurs when we need to associate data from different organizations (e.g., after a corporate merger, when combining information from two different divisions of the same company, etc.)—the databases are unlikely to share the same key. Another example occurs when the key information is entered manually, and with many typos. The technique for associating these tables without the benefit of an exact join key is to use an **approximate join** (or a **fuzzy join**).

A related problem occurs when imprecise information is used as the key of a table. For example, a customer's name might be used as the key of a customer account table. If the sales representatives manually enter these names, any given customer might be entered in the table with several creative misspellings of their name. The process of combining multiple entries into a single entry is called **duplicate elimination**.

Many vendors provide approximate join and duplicate elimination software and services. In this section, we briefly discuss the principles of how these systems work.

Approximate Field Matching

One of the fundamental tools for performing approximate joins is a method for determining that two things are similar, and also a method for computing how similar they are. For some data types, this is fairly easy. For example, if we are looking for a person who is about 40 years old, we can search our database for all people between the ages of 35 and 45. We would then order the output by the closeness of the listed age to 40.

Other data types require more sophisticated techniques, which we discuss below. Because so much of the data in a database consists of text, we pay particular attention to **string matching**.

String matching

To perform string matching, we need a metric which tells us how far apart two strings are. By analogy, we know that 38 is closer to 40 than 35 is because (40–38) is smaller than (40–35). The usual metric is the **string edit distance**, or how many elementary edit operations are required to change one string into another. The most common set of editing operations are (a) change any letter into any other, (b) insert any letter at any position, and (c) delete any letter. For example "Snark" is edit distance 2 from "Snort", but edit distance 5 from "Gryphon" (and therefore a Snark is closer to a Snort than to a Gryphon).

Other editing operations have been proposed to better capture typos (e.g., allow character transpose), but the basic three editing operations generally capture the meaning of "similar strings". There are many fast algorithms for computing the string edit distance, and they can be found in most elementary algorithms textbooks. Our problem is a little different, however. We want to compare a one string against a very large number of others, and to do so without making individual comparisons. (One string distance computation is fast; 100,000,000 of them is pretty slow.)

There are three basic approaches to finding all strings in a source list at distance k or less from a target string s:

1. Build an index on the source list of strings (this is easily accomplished when the source list is a field of a table in a database). Generate the set of all strings at edit distance k from s, and use the index to see if they are in the list. If k is larger than 1 or 2, this method will generate a very large number of index lookups.
2. Build a supplementary index on the target list as follows: For each string in the target list, compute the set of all q-grams, or q-character substrings. Create a supplementary table with two fields: a q-gram field and a source string field. Fill the supplementary table with an entry (Q,t) for every q-gram Q of every string t in the source list. To find source list strings which are close to target string s, compute the set of q-grams of s, and find a set of candidates from the source list strings whose set of q-grams is similar to the that of s. Compute the edit distance between s and all of the candidates, and return those at distance k or less.

   The idea behind the q-gram index approach is that any single string edit operation can affect at most q of the common q-grams of s and t. The effectiveness of the filter can be improved by storing not only the q-grams, but also their position in the string, and requiring not only that the q-grams match, but that their positions are close.
3. A hybrid of the above two approaches.

The second approach (q-gram index) has the advantage that the supplementary index table can be stored in the database, and the search for the candidate list is a simple SQL query which is readily optimized.

Tree matching

Some data, such as XML records, are tree structured (see Section 5.4). Because XML data are likely to be as dirty as relational data, it is sometimes valuable to perform approximate matching on tree data. The natural metric on tree structured data is the **tree edit distance**, analogous to the string edit distance. Unfortunately, computing the tree edit distance is very expensive (string edit distance is $O(n^2)$, where n is the number of characters, while tree edit distance is $O(l^4)$, where l is the number of leaves). Recent research has produced faster approximate algorithms, which we discuss in the annotated bibliography.

Feature vector matching

Similarity search is a basic data mining tool. The idea is to represent an object by a list of (usually numeric) features (the **feature vector**). The distance between two objects is usually the L_2 distance between their feature vectors. For example, a color can be represented by its reflectivity in the red, blue, and green spectrums. To find all colors similar to a particular shade of yellow, we look for colors with a close match on the red, blue, and green reflectivity. A large literature and many software tools have been developed for feature vector matching, and we provide some citations in the annotated bibliography.

Ad-hoc matching

In many cases, there is a special trick that can be exploited to find similar objects. For example, toll fraud investigators have found that calling patterns contain valuable information for uncovering instances of identity theft. Once they have determined that an account was taken out under false pretenses, they can save that calling pattern, since it contains the friends and relatives of the crook behind the identity theft. It is expected that the crooks will continue to call their friends and relatives even if they assume yet another identity. So by matching calling patterns from already compromised accounts to new accounts, fraud can be detected quickly, even with very dissimilar names on the accounts.

Approximate Joins and Duplicate Elimination

While approximate matching is a very useful tool for performing approximate joins, there is more to the story:

- Approximate matches are just that—approximate. To have confidence in the match, we need to look at **correlating information**. For example, if we match a customer in the billing database and the provisioning database, a customer who has a lot of equipment allocated to them should be generating large bills, and vice versa.
- Usually, no single field is always close. For example, if the customer name is alternatively recorded as "GE" and "General Eclectic", we are unlikely to approximately match these records on customer name. However if their phone numbers are the same, then we would have a reason to believe that they are the same entity.

Because of the application-specific nature of many approximate matching problems, many approximate-join software packages provide facilities for building an approximate-join application, rather than a pre-configured application (there are exceptions for very common approximate matching problems, e.g. address matching). Such a package typically works as follows:

- Define a set of buckets, and put the records from the tables to be matched in these buckets (in the case of duplicate elimination, there is only one table). There are several ways to define these buckets, including
  - Compute a hash function (i.e., a small summary) of one or more fields, sort the records on the hash, and use region around the sort position of the source record in the target list as the bucket for the source record.
  - Compute a hash function of one or more fields, and use an exact match on its value to define the buckets.
  - Use one of the approximate matching techniques to define the buckets.

  Records which reside in the same bucket are candidates for matching.
- For each pair of candidates, apply a match evaluation function. Accept the pairs with a high rating.

The exact working of the algorithm depend on the desired result. An approximate join algorithm matches candidates from the two different tables, while duplicate elimination matches records from the same table. The purpose of defining candidates as records which fall into the same bucket is to reduce the number of times that the (perhaps expensive) match evaluation function must be applied. If no prefiltering is done, the match evaluation function must be applied to every pair of records—$O(N^2)$ pairs, where N is the number of records. It is possible, even desirable, to use several bucket definitions to find match candidates (e.g., if the names are far apart, the telephone number might still be close). The hash functions might be very simple, for example, just the field value, or it might be a transformation of the field value. Common hash functions are Soundex (developed by the U.S. Census Bureau for duplicate elimination), and using the first few letters or consonants of a name.

The evaluation function is a heuristic rating of the similarity of two records. Because this function is run on a reduced set of candidate pairs, it can perform expensive operations, for example, look at a wide set of correlating information, perform multiple transformations to match fields, and so on. However, it is often the case that no single hash function will bring together matching records (e.g., sometimes it is the names that are similar, sometimes it is the telephone numbers). Fortunately this problem is easy to solve—use a set of hash functions and run the matching algorithm with each of them.

Thus the approximate-join software provides a matching engine which is customized by the user. The software will also provide libraries of

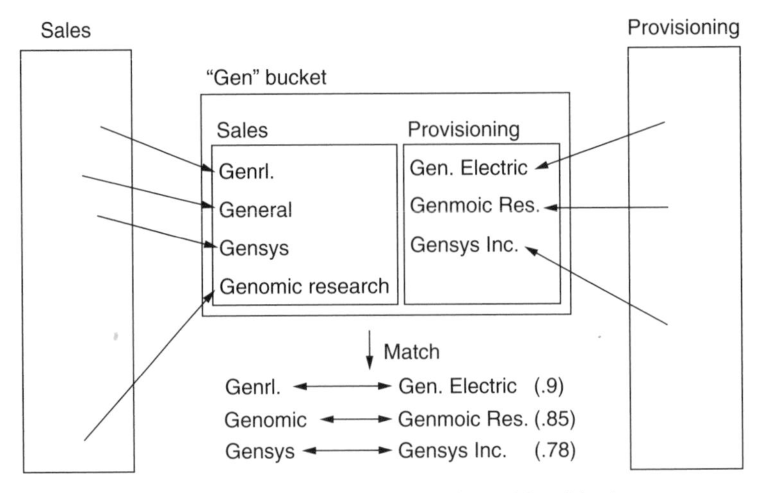

Figure 5.15: Approximate join of Sales and Provisioning.

transformation, hashing, approximate match, and match evaluation utilities. The matching and evaluation rules are often highly application specific, for example, does "St." expand to "Street" or "Saint". Exploratory data mining often helps to identify the best set of matching rules.

Let's consider an example, illustrated in Figure 5.15. We are trying to join data from two tables, Sales and Provisioning, using an approximate join. We the hash of a record to be the first three characters of the company name (a simple but probably not effective choice). Within the "Gen" bucket, we have four records from Sales and three records from Provisioning. The match evaluation function performs a variety of tests on the 12 combinations, and finds three pairs with a high rating (of .75 or higher).

When performing duplicate elimination, there is one additional consideration. Matching records refer to the same underlying entity. Therefore, if $R1$ matches to $R2$, and $R2$ matches to $R3$, we can infer that these three records all refer to the same entity, even if we would not otherwise match $R1$ to $R3$ (i.e., matching records form an **equivalence class**). More concretely, if record $R1$ with customer name "General Eclectic" matches to record $R2$ with customer name "GE" because of a common telephone number, it will also match to record $R3$ which has no fields in common, but matches to $R2$ on the customer name "GE". This matching tool is powerful but dangerous—it can greatly reduce the number of false negatives but greatly increase the number of false positives.

5.3.5 Database Profiling

In our experience, no data quality project can proceed without an extensive look at the data. Database profiling is just that—a systematic summarization of the contents of a database. Many of these summaries embody the EDM methods described in Chapters 2 and 3. For example, we have found the following query to be extremely helpful:

```
Select FIELD, count(*) as CNT
From TABLE
Group By FIELD
```

This SQL query reports the number of times each different value of field FIELD occurs in table TABLE. If we sort by the count in descending order ("Order By CNT DESC"), undocumented default values usually appear near the top of the list (we often find two or more). If we sort by the count in ascending order ("Order By CNT ASC"), we can often see typos. Viewing typical contents of a field generally reveals a wealth of unexpected insights.

In this section, we discuss some of the database-specific summaries that database profiling systems collect, and how they can be used. To a large extent, these summaries recover the database metadata from the data itself—a very useful function if the metadata is unavailable or untrustworthy.

Functional Dependencies and Keys

Recall from Section 5.3.1 that a well-designed database (i.e., normalized) has removed redundancies from its tables. In particular, there should be no **functional dependencies**, that is, a set of fields whose value determines the value of another field in the same table. For example, if we know the zip code of an address then we can determine the value of the state using only a lookup-table. Therefore the zip-code functionally determines the state[1], so any table which contains both zip code and state contains some degree of redundancy. A **key** is a set of fields which is unique in every record of a table. We note that a key can consist of more than one field. For example, a last name field is probably not unique in every record of a table, but a (first name, last name) combination might be. We are generally interested in minimal functional dependencies and keys (you can always add fields to a functional dependence or a key, and it is still a functional dependence/key). For example, if (first name, last name) is unique in every field, the (first name, last name, state) is also, but the smaller key is more useful. An **approximate functional dependency** is a set of fields which usually determine the value of another, and an **approximate key** is a set of fields which is unique in almost every record.

Fundamentally, the algorithms for finding keys and functional dependencies issue lots of counting queries to the DBMS. For example, if the number of unique values of the zip code is equal to the number of unique (zip code, state) pairs and there are more zip codes than states, then the zip code functionally determines the state in the table. Similarly, if the number of unique

[1]Actually, in some areas of sparsely populated western states, a zip code can span more than one state—it is an **approximate** functional dependency.

customer names is equal to the number of records in a table, then the customer name is a key for the table.

Key and functional dependence finding can be extremely expensive. If there are F fields in a table, then finding keys of up to k fields requires $O(F^k)$ expensive counting queries. Fortunately it is possible to reduce this cost, and we outline a fast key finding algorithm that we have found to be effective.

1. Eliminate all fields which are obviously bad candidates, for instance, because there are few unique values, because they are mostly null, because they have a data type such as floating point, and so forth.
2. Collect a sample of these fields in main memory.
3. Find keys within the sample in a level-wise fashion. That is, find all single-field keys, then two-field keys, and so on.
4. Issue count queries on the keys of the sampled table to verify that they are keys in the full table.

The point of finding keys in the sampled table is to allow the use of a pruning rule: don't bother computing the count of the number of unique values of a set of fields if some subset of these fields has already been determined to be a key or a functional dependency. Because the counting is done on a sample of the table, a sample key might not be an actual key, hence the verification step. However counting in the sampled table is much faster than in the full table, and serves as a very effective filter.

Finding functional dependencies in the tables of a database can help determine whether there is redundancy in the database schema design, possible leading to a more normalized database schema. However, there are a couple of caveats: (a) tables are often left un-normalized, to improve performance (i.e., avoid joins), and (b) there might be apparent but not actual functional dependencies in a table (e.g., zip codes appear to functionally determine state, but logically they do not). Perhaps more interesting are approximate functional dependencies. Why are they only approximate and not exact?

Finding keys and approximate keys helps us to find join paths, and to obtain correlating information for approximate joins. In many cases, we have been given data sets in which the field which is supposed to be the primary key was not actually unique in all records, but we were able to find other keys in the table and use them to join tables or to verify approximate joins.

Field Value Classification

In many data sets, the fields data types (or domains) are either not given, or are poorly specified. Field value classification attempts to recover these metadata. The simplest mechanisms classify fields by their length and by their character sets (numeric, alphabetical, etc.). More useful is a domain classification mechanism described in Section 5.3.3. We often know that a field contains text data of up to 20 characters, but does it appear to be a name, phone number, job title, or so on?

Join Paths

A **join path** is the sequence of one or more joins required to associate data in one table with the data in another. The join paths tell you the structure of a relational database—they indicate how information from different tables combines to describe an entity and its properties. Very often, this information is missing, incomplete, or inaccurate. However, we can recover much of this information by special field value classification mechanisms.

Finding join paths can be an extremely expensive task. A database of moderate complexity will have hundreds of tables and thousands of fields. If we try to find join paths by evaluating the join of all pairs of fields, we will submit $O(F^2)$ very expensive queries, where F is the number of fields in the database.

If we have classified fields into a set of domains, we can greatly reduce the number of field pairs that we need to test. That is, phone numbers will be matched to phone numbers, and so on. Still, a very large number of very expensive queries is needed.

There is a special sampling technique, called **min hash sampling**, which can help with finding join paths. A min hash sample of a field returns a small number (50 to 100) hash values, called the **signature** of a field. (A *hash* is a small summary of a value, usually an integer value.) A comparison of the signatures of two fields can compute their **resemblance**, or the size of the intersection of the set of unique values of the two fields divided by the size of the union. From the resemblance and the number of unique values in each of the two fields, we can compute the size of the intersection, and whether set of values of one field is contain in the other. This information is sufficient to find all pairs of fields which are likely to be joinable. Because the signatures are small and the resemblance computation is a simple query, finding these pairs is very fast.

In many cases, two fields can be joined after a small transformation. For example, one table might store customer names in the format "Lastname, Firstname", while the other stores it in the format "Firstname Lastname". While it is very difficult to automatically find join paths when transformations are required, it is possible to find fields which are "textually similar". Once pairs of textually similar fields are found, a quick visual inspection can usually determine whether the fields are potentially related and what transformations can be applied. The trick to finding textually related fields is to compute the min hash samples on the q-grams of the field values, rather than the field values themselves. Textually similar fields will have a largely overlapping set of q-grams, and therefore a high resemblance.

5.3.6 Annotated Bibliography

Good database textbooks include [34, 98]. These texts cover material on database design (keys, functional dependencies, normal forms, etc.) and the SQL query language. Rahm and Do [103] survey database-related research in on data quality issues.

Research in ETL has been performed in many guises, we provide an entry to the literature here. Access to legacy data is often assisted by the use of *mediators* [95]. Legacy or web data sources often provide limited query interfaces, requiring query transformations. This problem is addressed by [125], and Halevy [55] provides a survey of this technology. Even when the query interface does not present a problem, the source database often does not have the desired structure. Miller, Haas, and Hernandez [90] give algorithms for *schema mapping*. Lakshmanan, Sadri, and Subramanian [79] describe data transformation techniques such as pivot, unpivot, etc. Problems related to extracting data from free-form web sources are discussed in [10, 76]. Borkar, Deshmukh, and Sarawagi [11] present a method for segmenting free-form text into structured records. Rahm and Do [103] provide a summary of ETL research, and Vassiliadis et al. [126] describe an ETL system.

A summary of string matching algorithms can be found in [96, 97]. Gravano et al. [53] show how approximate string matching can be phrased as an SQL query. Koudas et al. [78] provide fast approximate tree (XML) matching algorithms. The book by Faloutsos [43] is an excellent source on (non-string) similarity search. The algorithms for approximate joins and duplicate elimination are from [58, 59, 94]. An industrial approximate join implementation is described in [16]. In [2], the authors propose using the hierarchical information commonly found in data warehouses. An alternative Artificial Intelligence based approach is described in [21].

An industrial implementation of database profiling is described in [120]. Efficient algorithms for finding keys and functional dependencies is given by [64]. Systems which summarize field values are described in [33, 104, 128]. A method for keyword search in a relational database is presented in [61].

Researchers have developed prototype data cleaning software systems. *Ajax* [49] provides extended database operators (for field transformation, approximate matching, and duplicate elimination) to allow the declarative expression of an ETL data flow. *Potter's Wheel* [104] is an ETL system with two notable features. First, it provides an interactive system for viewing the results of data transformations, making clear their effect. Second, it has a domain discovery mechanism, and uses it to find discrepancies in field values. Bellman [33] provides a suite of advanced profiling tools including min hash sampling.

5.4 METADATA AND DOMAIN EXPERTISE

In this section we give a brief overview of metadata and domain expertise, and techniques for creating and storing metadata. Consider the following scenario: A team of data quality gurus run extensive analyses and come up with a subset of records that are peculiar. The fields in these records behave very differently from the majority of the data set. The owners of the data are not impressed.

They say "Oh, we did that on purpose. Our GUI wouldn't allow us to add an additional field so we decided to put Service A and Service B together in the same field. We can tell the difference because Service A starts where Service B leaves off." The data quality gurus are understandably annoyed that this piece of information about the data was never mentioned to them, leave alone documented anywhere. In reality, a major part of data quality findings consist of unspecified and undocumented *metadata*, data about data.

An efficient way of recording, updating and transmitting metadata is to have a complete, accurate and up to date **schema**—essentially the metadata that is contained in a well-designed database that resides in a relational DBMS (see Section 5.3.1). Formal schemas come in three major flavors—Based on entity-relationships (**hierarchical schemas**), object-oriented schemas that include data encapsulation, and relational schemas as previously discussed.

Not many data sets come with good schemas. Although modern relational databases provide extensive facilities for schema documentation, they are often not used or not maintained. Also, in many instances the data set resides in legacy databases or in ad hoc structures. In these cases, the schemas are often informal and ad hoc, integrated into the programs that produce or use the data or in the heads of people, subject to mutation every time it is transferred from person to person. Worse, the knowledge is lost if the key person leaves without documenting the schema, making the data unusable. Most often the data reside in relational data bases, but when data are transmitted from one group to another, it often comes in delimited flat files, with a rudimentary schema describing the field layout.

When no formal schema exists, or the existing one is inadequate, the data users have to reverse engineer a schema from data (see Section 5.3.5). Data quality problems arise when the schema is incomplete or underspecified, or when the schema is ignored by the producer or the user of the data. Very often the schemas do not evolve as data changes. Even if the schema is updated, the data user has no way of knowing until she or he is notified. So the users often create their own mechanism for detecting and inferring changes in the schema. There are other sociological (reward sales but not documentation) and technical (limitations of systems, interfaces, expense of making system changes) issues that could result in improper or inadequate schemas.

A schema in the strictest definition of the word is sometimes not adequate. We need to know how to interpret values such as the scale that is used (sometimes values are "standardized" to make them more interpretable), units of measurement and meaning of labels for categorical attributes. Furthermore, we need to know how to interpret the tables—the frequency of refresh, associations between tables, the definition of the view that the tables represents Most work done in this areas has been for scientific databases where metadata can include programs for interpreting the data set (in an enterprise database, the metadata includes the business rules (dynamic constraints) for processing and interpreting the data). However, facilities for storing this metadata are readily available in relational databases, either directly (e.g., the

<tutorial>
<title> Problems, Solutions, & Research? <\title>
<Conference area="Statistics"> SIAM Data Mining Conference <\Conference>
<author> T. Dasu
<bio> Statistician <\bio> <\author>
<author> T. Johnson
<institution> AT&T Labs <\institution> <\author>
<\tutorial>

Figure 5.16: An example XML record.

stored procedures which interpret the data are part of the database itself), or indirectly (e.g., through user-created metadata tables). Unfortunately, these facilities are under-used and are often not transmitted with interchanged data.

In recent times, there has been interest in a more general platform for the exchange of metadata regarding data shared by multiple groups of users (such as a consortium of manufacturers of auto parts) and across multiple applications. Data interchange formats describe logical characteristics of shared data and contain semantics of multiple data-exchange applications. The data and the schema are intermingled and serve as a contract between multiple producers and consumers. Data exchange schemas are independent of storage system and data model making them more widely applicable. The close connection between schema and the data makes it harder to introduce data quality errors of misinterpretation by making it harder to circumvent the schema.

A data interchange format especially developed for web applications is XML. While XML is actually a text markup language (being derived from SGML, the Standard Generalized Markup Language), it has facilities and extensions for transmitting significant amounts of metadata. While a detailed discussion of such languages is outside the scope of this book, we give a brief description.

Figure 5.16 shows an example XML record. A quick glance at the record shows the "self-describing" nature of XML records—each field of the record is labeled with its name. Also note the nested structure of the record. The "tutorial" has a "title", a "Conference", and two "authors". One of the authors has a "bio", while the other has an "institution". The free-form tree structure of the data avoids data interchange problems of forcing data into an alien structure, instead the data can be presented in the manner in which it is recorded. For one example, there is no need to supply default values for fields, instead missing fields are literally missing. In addition to providing the names of the fields and their nested interrelations, fields can have *attributes*. For example, the field "Conference" has the attribute "area", whose value is "Statistics". Attributes provide a mechanism to provide additional information about the meaning of a field—that is, to provide additional metadata.

An XML record can be too free-form. For data interchange, we often want to specify a minimal set of fields which will be in any valid record. XML allows the specification of **Document Type Definitions**, or **DTDs**. The DTD of a record specifies a minimal set of fields in the record, and their nesting structure. The DTD of a record is specified as a URI in the document type declaration of the record. For example, the following document type declaration states that records rooted by "tutorial" have a DTD which can be found at http://www.research.att.com/~tamr/tutorial.dtd:

⟨!DOCTYPE tutorial SYSTEM "http://www.research.att.com/~tamr/tutorial.dtd"⟩

For complex data, the DTD has some limitations. One significant limitation is the lack of field domains (recall Section 5.3.1): we cannot specify that the text in a field is a number, a name, an address, or so forth. Furthermore, DTDs are not able to express many of the interesting constraints on the data set. To address these shortcomings, XML schema languages (such as the W3C XML Schema) have been proposed. Using XML Schema, the schema can declare that elements are of particular data types or domains.

5.4.1 Lineage Tracing

In data warehouses, the source data feeds are transformed, processed, cleaned, and aggregated to a great extent to make them usable and interpretable. However, this processing is often not well documented, making it difficult to understand how the values in a particular record in the data warehouse were derived. **Lineage tracing** records the process by which records in a table were derived—obviously a very useful piece of metadata. Lineage tracing can be done at the coarse level of tables or at a finer granularity of an individual record. The transformation steps can be recorded as a *graph* recording the transformation operators and data sources. The results of lineage tracing are useful for analysis (interpreting the probability distributions of truncated and censored attributes), debugging (why are there suspicious drops and spikes in the histograms of certain attributes) and for putting in place feedback loops. For example, if a customer disconnects a service (Attribute = status in Table = Customer is changed), make sure to de-allocate the associated resources (change Attribute = availability in Table = Circuit and Attribute = Action in Table = Sales). As a consequence, resources will be available for provisioning as soon as they are discontinued by a customer, avoiding the spurious "network saturation" problem.

5.4.2 Annotated Bibliography

A discussion of metadata issues can be found in [118]. In [114, 116], the authors propose the use of tags on data elements to attach metadata. An excellent book on XML is [57]. Additional XML resources are [26, 109]. Research literature on lineage tracing includes [25, 50].

5.5 MEASURING DATA QUALITY?

So, at the end of the day, how do we measure data quality? It is not difficult to define a set of numbers as a political exigency for all concerned parties to sign off on. Uniqueness, completeness, accuracy—choose any or all. We can ensure uniqueness by getting rid of seeming duplicates, only to find out later that they are differentiated by a minute variation in the representation of an attribute which is critical to a downstream application. Completeness can be induced by imputing values but might introduce a bias which can skew the results of the analysis at the end of the data chain. Accuracy cut-offs can be met by all kind of workarounds or simply playing with the definitions of the data. The biggest danger in adhering strongly to a mandated set of golden rules is that the data can be manipulated to meet these rules if there is enough political pressure on those responsible for those numbers.

Instead, the questions to ask are: "Do the data quality metrics truly reflect the quality of the data in terms of usefulness and reliability? Do the end users of the data see improvement in their applications and results when the data quality metrics improve?" In other words, are the metrics **directionally correct** with respect to the utility of the data to its consumers at various points in the data chain? To illustrate this, we discuss an example, which is based on a real case study. We have changed, for proprietary reasons, elements of the case study that will not impact the conclusions in any way. In some places we have simplified the processes for clarity of explanation.

5.5.1 Inventory Building—A Case Study

Holy Cow Corp. sells high-tech machinery to clients all across the United Stated in a highly competitive market. The machines come in many varieties and are assembled from different types of components in local warehouses geographically close to the client location. The company wants a complete and accurate inventory of its machinery (the ability to assemble a given type of machine) to be sold by its sales force. False positives (tell customer that Holy Cow can provide the machine when it cannot) and false negatives (tell customer Holy Cow cannot provide the machine to the customer when it has the capability) carry a very high penalty—the client will go to the competitor either way. The loss of every sale has a big impact on Holy Cow's ability to survive in a very tough market.

Task Description

A data quality task force was set up to complete the task of inventory building within an ambitious time period. The task force was backed by the top officers of the company.

The task force identified the databases that played a critical role in the problem of inventory building for the sales force. Many phone calls, meetings and e-mail exchanges were needed to accomplish this. The task force had to *use the name of the top management* on several occasions to escalate and get responses in a reasonable time frame. The following databases were identified:

- Operations DB (OPED)—Identifies the components that are available at each local warehouse, the latter identified by an address. For example

  Record 1:
    Warehouseid=XYZ;
    Machineid=XYZ1;
    Component A = 10;
    Component B = 1;
    Component C = 5;
    Address=180 park Avenue, Florham Park, NJ 07932;
    Machine = Type 1;
    Status = Deployed

  Record 2:
    Warehouseid=XYZ;
    Machineid=XYZ2;
    Component A = 6;
    Component B = 3;
    Component C = 5;
    Address=180 park Avenue, Florham Park, NJ 07932;
    Machine = Type 1;
    Status = Inactive

  Record 3:
    Warehouseid=XYZ;
    Machineid=XYZ1;
    Component A = 10;
    Component B = 1;
    Component C = 5;
    Address=180 Park Avenue Florham Park NJ 07932;
    Machine = Type 1;
    Status = Deployed

  where warehouse XYZ located at 180 Park Ave in Florham Park NJ has multiple records showing the type of machines, their status and the components used in building them. *DQ alert*: Records 1 and 3 are identical except for the commas in the address. In principle, the two records are duplicates and should be cleaned up. However, domain experts assure us that the duplication is necessary since some downstream databases can handle commas in the address and some cannot. The problem will hopefully be cleaned up when the downstream databases are fixed. However, there is a chance that they may never be cleaned up.

- Information Warehouse (IOWA)—A data warehouse that provides some additional information such as machine descriptions and the information about the owners.

  Record 1:

  Warehouseid=XYZ;
  Machineid=XYZ1;
  Component A = 10;
  Component B = 1;
  Component C = 5;
  Street Number=180;
  Street Name = Park Avenue;
  City=Florham Park;
  State=NJ;
  Zip=07932;
  Machine = Type 1;
  Status = Deployed;
  Machine Description= High powered reflector;
  Customer = Big Bad Wolf Inc.;
  Customer Contact = 973-555-1212

- Sales and Provisioning DB (SAPDB)—The information that is actually seen by the sales accountants who deal directly with the client to make the sale.

  Record 1:

  Warehouseid=XYZ;
  Machineid=XYZ1;
  Component A = 10;
  Component B = 1;
  Component C = 5;
  Street Number=180;
  Street Name = Park Avenue;
  City=Florham Park;
  State=NJ;
  Zip=07932;
  Machine = Type 1;
  Status = Deployed;

Machine Description= High powered reflector;

Customer = Big Bad Wolf Inc.;

Customer Contact = 973-555-1212;

Customer Address = 1500 Central Avenue, Madison, NJ;

When a sales person makes a sale, a request goes back to the field operations to assemble and deploy a particular type of machine at the customer's location.

The data flow consisted of:

OPED → IOWA → SAPDB

Metadata

The task force started receiving the data feeds after several weeks of negotiations, stalling and phone calls. The first task was browsing the data to answer two questions (a) what data flowed from OPED to SAPDB intact and (b) what data fell out or was changed in an unauthorized, unintended way while in transit via IOWA.

EDM performed on the three databases showed that:

- The documented metadata was a microscopic part of the metadata needed to correctly interpret the data—(a) Warehouseid in OPED was corrupted and there was a workaround for constructing the accurate Warehouseid that was not documented. So the initial efforts of matching the 3 DBs on Warehouseid failed.
- There were over 70 different machine types in OPED whereas only 10 were documented and defined.
- SAPDB database contained less than 15% of the records in OPED or IOWA.

Data quality alert: (a) The documentation is incomplete. (b) There is an intermediate step between IOWA and SAPDB that is causing a lot of data to fall out.

The taskforce went back to the data keepers with the discrepancies and found that (a) they did not know that many machine types existed, let alone what they mean (b) not all machines are made for sale to clients. Some machines are meant for the internal use of Holy Cow or to barter with competitors. The rules for what machine types are "sellable" and which are not were not clearly defined nor were they documented. Only the sellable machines were supposed to flow to SAPDB.

The taskforce then sought the help of a team of engineers to define each of the more than 70 machine types, the components needed to assemble them, the type of service and machine description and which were suitable for sale to customers and which were not. To build consensus on each of the definitions and establishing a standard and documenting it took more than 5 weeks of daily conference calls with participation ranging from 20–30 individuals from various departments. Finally, the rules were nailed down and the process of matching up the definitions and rules of flow against the data began.

Data Audits

The first data audit pass using custom built scripts that embody the definitions and rules i.e. business rules (dynamic constraints) revealed that (a) a big portion of the data did not flow to SAPDB because it was mislabeled with the wrong machine type and tagged as "internal use". (b) There were certain types of machines that were totally missing from OPED. (c) There were Machines that were duplicated in SAPDB with conflicting machine types.

A meeting with the data keepers explained that: (a) there were three separate manual workarounds by which data was entered into the system outside the process. There were thousands of such records where the machine type was manually changed to make them flow to SAPDB, causing the duplication. (b) There were small "databases" (more like informal spreadsheets) that existed in various groups across the country that did not make it into OPED, the alleged and intended DBoR (DataBase of Record). The satellite datasets were supposed to be entered into OPED but were not. (c) Furthermore, there were numerous users authorized to access OPED and change any data element. The taskforce found that there were many changes made daily without documentation or notification.

Several weeks were spent ensuring that the satellite data were included in OPED and that in the future would be entered directly into OPED and not into the spreadsheets. Explicit mandates to this effect from the highest officer in Holy Cow went out to all managers above a certain level. In the meantime, system fixes were put in place to avoid the manual workarounds.

By this time more than 60% of the allotted time had elapsed and the leader of the taskforce was getting seriously nervous. There was still no inventory, let alone complete or accurate.

The second pass indicated an improved flow from OPED to SAPDB, but the proportion of sales orders sent from SAPDB to field operations that were being thrown out as unprovisionable was still very high. A closer examination revealed that almost 15% to 20% of the records in OPED had conflicting addresses in SAPDB, resulting in the order being tagged unprovisionable since there was no means of determining the correct address of the local warehouse that contained the machine components.

Data quality alert: The data provided for the second audit run came in a different format then the feeds sent for the first audit run. This required either (a) the data should be re-sent in the correct format or (b) the scripts that performed the audit should be changed to incorporate the new format. The latter would require the scripts to be re-tested. In either case, the *changing format of the feeds* introduced delays.

An **address correction tool** was used to assign standardized addresses, but not before a pitted *political battle* was waged over several weeks to protest the interference with the systems. Despite the address correction, the third pass did not show any more improvement in the fallout rate between SAPDB and field operations. It turns out that OPED did not correctly reflect what was available and in place in local warehouses in certain regions, especially the Northeast. A painstaking **manual check** of each and every warehouse that had at least one fallout was commissioned to be completed within 10 days—a team of over 15 people worked literally 16–18-hour days to achieve this task. Again, such dedication and compliance came about primarily due to the mandate of the top ranking managers in Holy Cow. OPED was corrected to reflect the accurate data in the warehouse.

The Accomplishment

The final audit pass (which was again plagued by changing feed formats), conducted two days before the deadline, showed a vast improvement.

- The unintended fall out rate (as opposed to records that were designed to fall out) between OPED and SAPDB via IOWA had fallen to 2% from an incredible 50%–60%;
- OPED's inventory had swelled by over 15% (translates to tens of thousands of records) due to the inclusion of all the satellite databases;
- Over 10 system fixes had been put in place to improve automation and avoid manual entries;
- Addresses were standardized;
- Almost 98% of the data were flowing correctly as defined from OPED to IOWA to SAPDB, up from less than 40%. As a consequence, the inventory of sellable machines increased dramatically;
- The taskforce had mended over 70% of the data one way or another;
- The biggest achievement was a clear documentation of the rules and conventions that could serve as reference and a starting point for keeping track of the system as it evolves in the future.

Measuring Data Quality

In Section 4.5, we describe in general how constraints can be used as data quality metrics. What about in this specific case? There are several metrics we could use that reflect the utility of the data clean up efforts to the users and the systems. Some significant metrics are listed below, with parenthetic references to the general components from Section 4.5 that they correspond to.

- Increase in usable data, up from approximately 40% to 98%, as measured by conformance to schema (static metric) and conformance to business rules (dynamic metric). As a side effect of listing and testing the business rules, we actually found problems with the business process and managed to resolve them.

- Consolidation of data into one single source, the DBoR OPED, eliminating satellite databases that needed to be fed in manually (extent of automation—a dynamic, operational metric; completeness, a diagnostic metric);
- Increase in "sellable" inventory (diagnostic metric of accuracy i.e., usability, entails validation of static constraints);
- Reduction in manual workarounds (extent of automation—dynamic, operational DQ metric);
- Increase in accurate flow, where accurate implies "working as per specifications" (successful completion of end-to-end process—operational metric involving the satisfaction of dynamic constraints);
- A clear documentation of the data definitions, constraints, rules and specifications (diagnostic metrics of accessibility and interpretability).

The list is incomplete since there are many intangibles that cannot be captured by metrics. However, the point is that the metrics themselves are highly problem-specific and should be tailored to the business problem that is being solved by the data quality task, in consultation with the data creators, data keepers, data users and the subject matter experts.

5.5.2 Learning and Recommendations

What did we learn from the case study? What recommendations can we make?

- **Learning:**
  Take NOTHING for granted. The data are never what they are supposed to be, even after they are "cleaned up". The schemas, layouts, content and nature of content are never completely known or documented and continue to change dynamically.
  Recommendation(1):
  Check and re-check schema constraints and business rules (dynamic constraints) (*constraint checks*) every time fresh data arrive. Immediately identify and send discrepancies to responsible parties. (*Feedback loop*).
  Recommendation(2):
  Maintain a good relationship with the data owners and data creators to keep up with the changes as well to ensure a quick turnaround in case of problems.
- **Learning:**
  A big ingredient of the success of a data quality effort (e.g., directed at cleaning up operations) is the backing of upper management. The best way to convince managers about the importance of a data quality effort (or any

undertaking) is to quantify the impact in terms of dollars. It was noted that the lack of an accurate inventory was costing Holy Cow approximately one billion dollars annually in terms of various charges, fees and settlements. The taskforce in the case study would have not succeeded to the extent it did as well as in that short a time frame (3 months) without the mandate from the top management of Holy Cow, which in turn was motivated by the big dollar impact. The taskforce invoked the dollar mantra and the name of the higher-ups several times during its three-month mission.

Recommendation:

Line up big management names who are willing to intervene in case of uncooperative partners. Given that data are a power source, such issues of data sharing and knowledge sharing arise very frequently.

- **Learning:**

  A major source of data quality errors is manual entry and manual intervention.

  Recommendation(1):

  Data entry as well as other data processes should be *fully automated* in such a manner that data need be entered only once. Furthermore, data should only be entered and processed as per schema and business specifications.

  Recommendation(2):

  Perform *continuous and end-to-end audits* to immediately identify discrepancies. In fact, these should be a routine part of data processing.

- **Learning:**

  Lack of a clear specification of the data schema (static constraints) and business rules (dynamic constraints) are a major obstacle to the usability of the data. The Holy Cow taskforce spent a major chunk of its time acquiring schemas and researching business rules.

  Recommendation:

  Maintain an updated and accurate view of the schema and business rules. Use proper software and tools such as XML to enable this.

- **Learning:**

  A primary reason for the sad state of Holy Cow's data processes was that each separate organization was looking after its own piece and no one had the overall responsibility for the entire process.

  Recommendation(1):

  Appoint a *data steward* who owns the entire process and is accountable for the quality of the data. Reward "good" data.

  Recommendation(2):

  Publish the data where it can be seen and used by as many users as possible so that data discrepancies are more likely to be reported. Too often data are cloaked in secrecy and protected by bureaucratic procedures (submit a work request, specify business need, get authorization from manager X, can only see a few tables, cannot issue queries, and so on).

5.6 DATA QUALITY AND ITS CHALLENGES

In this chapter and the previous one, we have discussed data quality's multi-faceted nature. Data quality cannot be captured with a set of numbers or static rules. We must design processes, systems and analyses explicitly to monitor, isolate, and repair data that do not conform to a set of rules. These, in turn, are constantly scrutinized, updated, and documented as the data quality needs change. An additional challenge in measuring data quality is the sheer complexity of the problem given that there are more exceptions than rules while constructing a description of what constitutes good data. The set of rules is ever changing, often incomplete (insufficient metadata) and a significant, challenging data set in its own right! Furthermore, the rules are highly context specific, and need to be put together on a case-by-case basis. Therefore, creating a general, reusable solution that can automatically scan data and create a set of data quality constraints, and isolate the records that do not meet this process is quite far from reality at this point. It is a painstakingly manual process, involving many iterations, validity checks, and interaction between domain experts, data creators, data keepers, and data users.

We have to focus particularly on metadata and domain expertise in data quality projects, since they determine the success of a cleanup effort to a large extent. Data mining and browsing can provide additional insights and fill gaps in metadata.

Tools and algorithms in this area are in their infancy right now. Since the problem is complex and difficult to automate, existing algorithms attack small portions of the problem. A data quality implementer has to use a combination of such tools (Bellman, AJAX, Potter's wheel) and commercially available software to automate and speed up wherever possible.

In the end, the best defense is relentless monitoring of data and metadata, continuous auditing and validating the data/metadata against domain expertise using tools, algorithms and old-fashioned "looking over the data" to see if it all makes sense.

Bibliography

[1] F. Alt and N. Smith. Multivariate process control. In *Handbook of Statistics*, *7*, pages 333–351. North Holland, New York, 1988.

[2] R. Ananthakrishna, S. Chaudhuri, and V. Ganti. Eliminating fuzzy duplicates in data warehouses. In *Intl. Conf. Very Large Databases*, 2002.

[3] H. B. Aradhye, B. R. Bakshi, R. A. Strauss, and J. F. Davis. Multiscale statistical process control using wavelets. Technical report, Ohio State University, 2000.

[4] M. Barnsley and A. D. Sloan. A better way to compress images. *Byte*, pages 215–223, January 1988.

[5] D. Belsley, E. Kuh, and R. E. Welsch. *Regression Diagnostics*. Wiley, New York, 1980.

[6] A. Belussi and C. Faloutsos. Estimating the selectivity of spatial queries using the "correlation" fractal dimension. In *Intl. Conf. Very Large Databases*, pages 299–310, 1995.

[7] S. Berchtold, C. Bohm, and H.-P. Kriegel. The pyramid-tree: Breaking the curse of dimensionality. In *Proc. ACM SIGMOD Conf.*, pages 142–153, 1998.

[8] P. Billingsley. *Probability and Measure*. Wiley, New York, 1995.

[9] C. M. Bishop. *Neural Networks for Pattern Recognition*. Oxford University Press, Oxford, 1995.

[10] V. Borkar, K. Deshmukh, and S. Sarawagi. Automatically extracting structure from free text addresses. *Data Engineering Bulletin*, 23(4):27–32, 2000.

[11] V. R. Borkar, K. Deshmukh, and S. Sarawagi. Automatic segmentation of text into structured records. In *Proc. ACM SIGMOD Conf.*, 2001.

[12] M. M. Breunig, H.-P. Kriegel, R. T. Ng, and J. Sander. LOF: Identifying density-based local outliers. In *Proc. ACM SIGMOD Conf.*, pages 93–104, 2000.

[13] Cap Gemini Ernst & Young. Integrated revenue assurance for regulatory compliance & increasing profits. http://www.cgey.com/tmn/nmi/revenue_assurance/assets/Global%20Revenue%20Assurance%20flyer%20FINAL.pdf.

[14] D. Carlson. Data stewardship in action. *DM Review*, May 2002.

[15] L. Carroll. *Alice's Adventures in Wonderland*. Macmillan, New York, 1877.

Exploratory Data Mining and Data Cleaning, by Tamraparni Dasu and Theodore Johnson
ISBN: 0-471-26851-8

[16] F. Caruso, M. Cochinwala, U. Ganapathy, G. Lalk, and P. Missier. Telcordia's database reconciliation and data quality analysis tool. In *Intl. Conf. Very Large Databases*, pages 615–618, 2000.

[17] Chartered Accountants of Canada. Continuous auditing. Technical report, Canadian Institute of Chartered Accountants (CICA), Toronto, ON, 1999. http://www.cica.ca/cica/cicawebsite.nsf/public/sgrscontinuousauditing.

[18] S. Chaudhuri and U. Dayal. On overview of data warehousing and OLAP technology. *ACM SIGMOD Record*, 26(1):65–74, 1997.

[19] P. Cheeseman and J. Stutz. Bayesian classification (autoclass): Theory and results. In *Advances in Knowledge Discovery and Data Mining*, 1995.

[20] R. C. H. Cheng and T. C. Iles. Confidence bands for cumulative distribution functions of continuous random variables. *Technometrics*, 25:77–86, 1983.

[21] W. Cohen, H. A. Kautz, and D. A. McAllester. Hardening soft information sources. In *Intl. Conf. Knowledge Discovery and Data Mining*, pages 255–259, 2000.

[22] T. M. Cover and J. A. Thomas. *Elements of Information Theory*. Wiley, New York, 1991.

[23] D. R. Cox and D. V. Hinkley. *Theoretical Statistics*. Wiley, New York, 1974.

[24] D. R. Cox and D. Oakes. *Analysis of Survival Data*. Chapman and Hall, London, 1984.

[25] Y. Cui and J. Widom. Lineage tracing for general data warehouse transformations. In *Intl. Conf. Very Large Databases*, pages 471–480, 2001.

[26] J. S. D. Florescu. XML data: From research to standards. VLDB 2000 Tutorial, http://www.db.research.bell-labs.com/user/simeon/vldb2000.ppt, 2000.

[27] R. D'Agostino and M. Stephens. *Goodness-of-Fit Techniques*. Marcel Dekker, New York, 1986.

[28] S. Dasgupta. Learning mixtures of Gaussians. In *IEEE Symp. Foundations of Computer Science*, pages 634–644, 1999.

[29] T. Dasu and T. Johnson. An efficient method for representing, analyzing and visualizing massive high dimensional data sets. *Computing Sciences and Statistics*, 29, 1997.

[30] T. Dasu and T. Johnson. Hunting of the Snark: Finding data glitches using data mining methods. In *MIT Workshop on Information Quality*, pages 89–98, 1999.

[31] T. Dasu and T. Johnson. Approximating nonlinear models. In *15th International Workshop on Statistical Modelling*, 2000.

[32] T. Dasu, T. Johnson, and E. Koutsofios. Hunting data glitches in massive time series data. In *MIT Workshop on Information Quality*, 2000.

[33] T. Dasu, T. Johnson, S. Muthukrishnan, and V. Shkapenyuk. Mining database structure; or, how to build a data quality browser. In *Proc. ACM SIGMOD Conf.*, 2002.

[34] D. Date. *An Introduction to Database Systems*. Addison Wesley, Reading, MA, 1995.

[35] A. Dempster, N. Laird, and D. Rubin. Maximum likelihood from incomplete data via the EM algorithm. *Journal of the Royal Statistical Society*, *B*, 39:1–38, 1977.

[36] P. Domingo and M. Pazzani. Beyond independence: conditions for optimality of the simple bayesian classifier. In *Intl. Conf. Machine Learning*, pages 105–112, 1996.

[37] D. Donoho. *Breakdown properties of multivariate location estimators*. PhD thesis, Harvard University, 1982.

[38] R. Duda and P. Hart. *Pattern Classification and Scene Analysis.* Wiley, New York, 1973.

[39] W. DuMouchel, C. Volinsky, T. Johnson, C. Cortez, and D. Pregibon. Squashing flat files flatter. In *Intl. Conf. Knowledge Discovery and Data Mining*, pages 6–16, 1999.

[40] A. J. Duncan. *Quality Control and Industrial Statistics.* Irwin, Homewood, 1974.

[41] L. English. *Improving Data Warehouse and Business Information Quality: Methods for Reducing Costs and Increasing Profits.* Wiley, New York, 1999.

[42] A. Eubank. *Spline Smoothing and Nonparametric Regression.* Marcel Dekker, New York, 1988.

[43] C. Faloutsos. *Searching Multimedia Databases By Content.* Klewer, Boston, 1996.

[44] W. Feller. *An Introduction to Probability Theory and its Applications, vol. 1.* Wiley, New York, 1968.

[45] W. Feller. *An Introduction to Probability Theory and its Applications, vol. 2.* Wiley, New York, 1968.

[46] R. A. Fisher. *The Design of Experiments.* Hafner (Macmillan), New York, 1966.

[47] D. Freedman, R. Pisani, and R. Purves. *Statistics.* Norton, New York, 1978.

[48] J. Friedman and B. Silverman. Flexible parsimonious smoothing and additive modeling. *Technometrics*, 31:3–39, 1989.

[49] H. Galhardas, D. Florescu, D. Shasha, E. Simon, and C.-A. Saita. Declarative data cleaning: Language, model, and algorithms. In *Intl. Conf. Very Large Databases*, pages 371–380, 2001.

[50] H. Galhardas, D. Florescu, D. Shasha, E. Simon, and C.-A. Saita. Improving data cleaning quality using a data lineage facility. In *Proc. 3rd Intl. Workshop Design and Management of Data Warehouses,* pages 3-1–3-10, 2001.

[51] V. Ganti, R. Ramakrishnan, J. Gehrke, A. L. Powell, and J. C. French. Clustering large datasets in arbitrary metric spaces. In *IEEE Intl. Conf. Data Engineering*, pages 502–511, 1999.

[52] W. G. Gilchrist. *Statistical Modelling With Quantile Functions.* Chapman & Hall/CRC, Boca Raton, FL, 2000.

[53] L. Gravano, P. Ipeirotis, H. Jagadish, N. Koudas, S. Muthukrishnan, and D. Srivastava. Approximate string joins in a database (almost) for free. In *Intl. Conf. Very Large Data Bases*, pages 491–500, 2001.

[54] J. Gray, A. Bosworth, A. Layman, and H. Pirahesh. Data cube: A relational aggregation operator generalizing group-by, cross-tab, and sub-total. In *IEEE Intl. Conf. Data Engineering*, pages 152–159, 1996.

[55] A. Halevy. Answering queries using views: A survey. *VLDB Journal*, 10(4):270–294, 2001.

[56] J. Han and M. Kamber. *Data Mining: Concepts and Techniques.* Morgan Kaufmann, San Francisco, 2000.

[57] E. Harold and W. Means. *XML in a Nutshell, a Desktop Quick Reference.* O'Reilly, Sebastopol, CA, 2001.

[58] M. Hernandez and S. Stolfo. The merge/purge problem for large databases. In *Proc. ACM SIGMOD Conf.*, pages 127–135, 1995.

[59] M. Hernandez and S. Stolfo. Real-world data is dirty: Data cleansing and the merge/purge problem. *Data Mining and Knowledge Discovery*, 2(1):9–37, 1998.

[60] D. C. Hoaglin, F. Mosteller, and J. Tukey. *Understanding Robust and Exploratory Data Analysis.* Wiley, New York, 1983.

[61] V. Hristidis and Y. Papakonstantinou. DISCOVER: Keyword search in relational databases. In *Intl. Conf. Very Large Databases*, 2002.

[62] K.-T. Huang. Y. W. Lee. and R. Y. Wang. *Quality Information and Knowledge Management.* Prentice Hall, New Jersey, 1999.

[63] P. J. Huber. *Robust Statistics.* Wiley, New York, 1981.

[64] Y. Huhtala, P. Porkka, and H. Toivonen. Tane: An efficient algorithm for discovering functional and approximate dependencies. *The Computer Journal*, 42(2):100–111, 1999.

[65] S. I. Inc. SAS STAT user's guide, version 6, fourth edition, volume 1, 1989. Cary, NC: SAS Institute Inc.

[66] S. I. Inc. SAS STAT user's guide, version 6, fourth edition, volume 2, 1989. Cary, NC: SAS Institute Inc.

[67] H. V. Jagadish, N. Koudas, S. Muthukrishnan, V. Poosala, K. C. Sevcik, and T. Suel. Optimal histograms with quality guarantees. In *Intl. Conf. Very Large Databases*, pages 275–286, 1998.

[68] A. K. Jain and R. C. Dubes. *Algorithms for Clustering.* Prentice Hall, New Jersey, 1988.

[69] A. K. Jain, M. N. Murty, and P. J. Flynn. Data clustering: a review. *ACM Computing Surveys*, 31(3):264–323, 1999.

[70] B. J. Jansen, A. Spink, and T. Saracevic. Real life, real users, and real needs: a study and analysis of user queries on the web. *Information Processing and Management*, 36(2):207–227, 2000.

[71] T. Johnson and T. Dasu. Comparing massive high-dimensional data sets. In *Knowledge Discovery and Data Mining*, pages 229–233, 1998.

[72] T. Johnson and T. Dasu. Scalable data space partitioning in high dimensions. In *The Statistical Computing Section, American Statistical Association*, 1999.

[73] T. Johnson, I. Kwok, and R. Ng. Fast computation of 2-dimensional depth contours. In *Intl. Conf. Knowledge Discovery and Data Mining*, pages 224–228, 1998.

[74] J. Kalbfeisch and R. Prentice. *The statistical analysis of failure time data.* Wiley, New York, 1980.

[75] J. P. Klein and M. L. Moeschberger. *Survival Analysis Techniques for Censored and Truncated Data.* Springer-Verlag, New York, 1997.

[76] C. Knoblock, K. Lerman, S. Minton, and I. Muslea. Accurately and reliably extracting data from the web. *Data Engineering Bulletin*, 23(4):33–41, 2000.

[77] E. Knorr and R. Ng. Algorithms for mining distance-based outliers in large datasets. In *Proc. Intl. Conf. Very Large Data Bases*, pages 392–403, 1998.

[78] N. Koudas, D. Srivastava, H. Jagadish, S. Guha, and T. Yu. Approximate XML joins. In *Proc, ACM SIGMOD Conf.*, 2002.

[79] L. Lakshmanan, F. Sadri, and S. Subramanian. SchemaSQL: An extension to SQL for multidatabase interoperability. *ACM Trans. On Database Systems*, 26(4):476–519, 2001.

[80] E. L. Lehmann. *Nonparametrics: Statistical Methods Based on Ranks.* Holden Day, Oakland, California, 1975.

[81] R. J. A. Little and D. B. Rubin. *Statistical Analysis with Missing Data.* Wiley, New York, 1987.

[82] R. Liu, J. Parelius, and K. Singh. Multivariate analysis by data depth: descriptive statistics. *Ann. Statist.*, 27:783–858, 1999.

[83] R. Liu and K. Singh. A quality index based on data depth and multivariate rank tests. *J. Amer. Statist. Assoc.*, 88:252–260, 1993.

[84] D. Loshin. *Enterprise Knowledge Management: The Data Quality Approach.* Morgan Kaufmann, San Francisco, 2001.

[85] G. S. Manku, S. Rajagopalan, and B. G. Lindsay. Approximate medians and other quantiles in one pass and with limited memory. In *Proc. ACM SIGMOD Conf.*, pages 426–435, 1999.

[86] P. McCullagh and J. A. Nelder. *Generalized Linear Models.* Chapman and Hall, London, 1989.

[87] G. McGraw and J. Viega. Make your software behave: Learning the basics of buffer overflows, 2000. http://www-4.ibm.com/software/developer/library/overflows/index.html.

[88] K. Miller, S. Ramaswami, P. Rousseeuw, T. Sellares, D. L. Souvaine, I. Streinu, and A. Struyf. Fast implementation of depth contours using topological sweep. In *Proc. Symp. Discrete Algorithms*, pages 690–699, 2001.

[89] K. Miller, S. Ramaswami, P. J. Rousseeuw, T. Selares, D. Souvaine, I. Streinu, and A. Struyf. Efficient computation of location depth contours by methods of combinatorial geometry. Technical report, University of Antwerp, 2001.

[90] R. Miller, L. Haas, and M. Hernandez. Schema mapping as query discovery. In *Intl. Conf. Very Large Databases*, pages 77–88, 2000.

[91] M. C. Minnotte. Achieving higher-order convergence rates for density estimation with binned data. *J. American Statistical Association*, 93(442):663–672, 1998.

[92] M. C. Minnotte, D. J. Marchette, E. J. Wegman. The bumpy road to the mode forest. *J. Computational and Graphical Statistics*, 7(2):239–251, 1998.

[93] D. Mladenic and M. Grobelnik. Text mining: What if your data is made of words?, 2001. Tutorial at ECML/PKDD-2001.

[94] A. Monge. Matching algorithms within a duplicate detection system. Data *Engineering Bulletin*, 23(4):14–20, 2000.

[95] P. S. M. T. Roth. Don't scrap it, wrap it! a wrapper architecture for legacy data sources. In *Intl. Conf. Very Large Databases*, pages 266–275, 1997.

[96] G. Navarro. A guided tour to approximate string matching. *ACM Computer Surveys*, 33(1):31–88, 2001.

[97] G. Navarroa, R. Beaza-Yates, E. Sutinen, and J. Tarhio. Indexing methods for approximate string matching. *Data Engineering Bulletin*, 24:19–27, 2001.

[98] P. O'Neil. *Database Principles Programming Performance.* Morgan Kaufmann, San Francisco, 1994.

[99] J. Pearl. Bayesian networks. Technical Report 980002, Computer Science dept., UCLA, 31, 1998. citeseer.nj.nec.com/pearl00bayesian.html.

[100] R. K. Pearson. Data mining in the face of contaminated and incomplete records. In *SIAM Intl. Conf. Data Mining*, 2002.

[101] F. P. Preparata and M. I. Shamos. *Computational Geometry: An Introduction.* Springer-Verlag, Berlin, 1985.

[102] D. Pyle. *Data Preparation for Data Mining*. Morgan Kaufmann, San Francisco, 1999.

[103] E. Rahm and H. Do. Data cleaning: Problems and current approaches. *Data Engineering Bulletin*, 23(4):3–13, 2000.

[104] V. Raman and J. Hellerstein. Potters wheel: An interactive data cleaning system. In *Intl. Conf. Very Large Databases*, pages 381–390, 2001.

[105] C. R. Rao. *Linear Statistical Inference and Its Applications.* Wiley, New York, 1973.

[106] T. Redman. *Data Quality: Management and Technology.* Bantam Books, New York, 1992.

[107] T. Redman. *Data Quality: The Field Guide.* Digital Press (Elsevier), 2001.

[108] P. R. Rosenbaum and D. B. Rubin. The central role of the propensity score in observational studies for causal effects. *Biometrika*, 70:41–55, 1983.

[109] A. Rosenthal. XML's impact on databases and data sharing. *IEEE Computer*, pages 59–67, 2000.

[110] S. Ross. *A First Course in Probability*. Macmillan, New York, 1988.

[111] P. Rousseeuw and A. Struyf. Computing location depth and regression depth in higher dimensions. *Statistics and Computing*, 8:193–203, 1998.

[112] P. J. Rousseeuw and M. Hubert. Regression depth. *Journal of the American Statistical Association*, 94:388–402, 1999.

[113] J. L. Schafer. *Analysis of Incomplete Multivariate Data.* Chapman & Hall, London, 1997.

[114] E. Sciore, M. Siegel, and A. Rosenthal. Using semantic values to facilitate interoperability among heterogenous information sources. *ACM Trans. On Database Systems*, 19(2):255–190, 1994.

[115] D. W. Scott. *Multivariate Density Estimation.* Wiley, New York, 1992.

[116] L. Seligman and A. Rosenthal. A metadata resource to promote data integration. In *IEEE Metadata Workshop*, 1996.

[117] R. J. Serfling. *Approximation Theorems of Mathematical Statistics.* Wiley, New York, 1980.

[118] G. C. Simsion. *Data Modeling Essentials, 2nd Edition: A Comprehensive Guide to Data Analysis, Design, and Innovation.* Coriolis Group, 2001.

[119] A. Singhal. Modern information retrieval: A brief overview. *IEEE Data Engineering Bulletin*, 24(4):35–43, 2001.

[120] E. Software. Data profiling and mapping, the essential first step in data migration and integration projects. http://www.evokesoftware.com/pdf/wtpprDPM.pdf.

[121] A. Struyf and P. J. Rousseeuw. High-dimensional computation of the deepest location. *Computational Statistics and Data Analysis*, 34:415–426, 2000.

[122] The Data Warehouse Institute. Data quality and the bottom line: Achieving business success through a commitment to high quality data. commissioned by Data Flux.

[123] J. Tukey. *Exploratory Data Analysis.* Addison-Wesley, Reading, 1977.

[124] J. W. Tukey. Nonparametric estimation, III. statistically equivalent blocks and multivariate tolerance regions—the discontinuous case. *Ann. Math. Stat.*, 27:783–858, 1948.

[125] V. Vassalos and Y. Papakonstantinou. Describing and using query capabilities of heterogeneous sources. In *Intl. Conf. Very Large Databases*, pages 256–265, 1997.

[126] P. Vassiliadis, Z. Vagena, S. Skiadopoulos, N. Karayannidis, and T. Sellis. Arktos: A tool for data cleaning and transformation in data warehouse environments. *Data Engineering Bulletin*, 23(4):42–47, 2000.

[127] R. L. Wang. *Journey to Data Quality*, volume 23 of *Advances in Database Systems*. Kluwer, Boston, 2002.

[128] L.-L. Yan, R. Miller, L. Haas, and R. Fagin. Data-driven understanding and refinement of schema mappings. In *Proc. ACM SIGMOD Conf.*, 2001.

[129] Y. C. Yang. Multiple imputation for missing data: Concepts and new development. *SAS Institute, Inc.*, 2000.

[130] Y. Zhu and D. Shasha. Statstream: Monitoring thousands of high speed data streams. In *Intl. Conf. Very Large Databases*, 2002.

Index

WILEY SERIES IN PROBABILITY AND STATISTICS

ESTABLISHED BY WALTER A. SHEWHART AND SAMUEL S. WILKS

The ***Wiley Series in Probability and Statistics*** is well established and authoritative. It covers many topics of current research interest in both pure and applied statistics and probability theory. Written by leading statisticians and institutions, the titles span both state-of-the-art developments in the field and classical methods.

Reflecting the wide range of current research in statistics, the series encompasses applied, methodological and theoretical statistics, ranging from applications and new techniques made possible by advances in computerized practice to rigorous treatment of theoretical approaches.

This series provides essential and invaluable reading for all statisticians, whether in academia, industry, government, or research.

ABRAHAM and LEDOLTER · Statistical Methods for Forecasting
AGRESTI · Analysis of Ordinal Categorical Data
AGRESTI · An Introduction to Categorical Data Analysis
AGRESTI · Categorical Data Analysis, *Second Edition*
ANDĚL · Mathematics of Chance
ANDERSON · An Introduction to Multivariate Statistical Analysis, *Second Edition*
*ANDERSON · The Statistical Analysis of Time Series
ANDERSON, AUQUIER, HAUCK, OAKES, VANDAELE, and WEISBERG · Statistical Methods for Comparative Studies
ANDERSON and LOYNES · The Teaching of Practical Statistics
ARMITAGE and DAVID (editors) · Advances in Biometry
ARNOLD, BALAKRISHNAN, and NAGARAJA · Records
*ARTHANARI and DODGE · Mathematical Programming in Statistics
*BAILEY · The Elements of Stochastic Processes with Applications to the Natural Sciences
BALAKRISHNAN and KOUTRAS · Runs and Scans with Applications
BARNETT · Comparative Statistical Inference, *Third Edition*
BARNETT and LEWIS · Outliers in Statistical Data, *Third Edition*
BARTOSZYNSKI and NIEWIADOMSKA-BUGAJ · Probability and Statistical Inference
BASILEVSKY · Statistical Factor Analysis and Related Methods: Theory and Applications
BASU and RIGDON · Statistical Methods for the Reliability of Repairable Systems
BATES and WATTS · Nonlinear Regression Analysis and Its Applications
BECHHOFER, SANTNER, and GOLDSMAN · Design and Analysis of Experiments for Statistical Selection, Screening, and Multiple Comparisons
BELSLEY · Conditioning Diagnostics: Collinearity and Weak Data in Regression

*Now available in a lower priced paperback edition in the Wiley Classics Library.

BELSLEY, KUH, and WELSCH · Regression Diagnostics: Identifying Influential Data and Sources of Collinearity
BENDAT and PIERSOL · Random Data: Analysis and Measurement Procedures, *Third Edition*
BERRY, CHALONER, and GEWEKE · Bayesian Analysis in Statistics and Econometrics: Essays in Honor of Arnold Zellner
BERNARDO and SMITH · Bayesian Theory
BHAT and MILLER · Elements of Applied Stochastic Processes, *Third Edition*
BHATTACHARYA and JOHNSON · Statistical Concepts and Methods
BHATTACHARYA and WAYMIRE · Stochastic Processes with Applications
BILLINGSLEY · Convergence of Probability Measures, *Second Edition*
BILLINGSLEY · Probability and Measure, *Third Edition*
BIRKES and DODGE · Alternative Methods of Regression
BLISCHKE AND MURTHY (editors) · Case Studies in Reliability and Maintenance
BLISCHKE AND MURTHY · Reliability: Modeling, Prediction, and Optimization
BLOOMFIELD · Fourier Analysis of Time Series: An Introduction, *Second Edition*
BOLLEN · Structural Equations with Latent Variables
BOROVKOV · Ergodicity and Stability of Stochastic Processes
BOULEAU · Numerical Methods for Stochastic Processes
BOX · Bayesian Inference in Statistical Analysis
BOX · R. A. Fisher, the Life of a Scientist
BOX and DRAPER · Empirical Model-Building and Response Surfaces
*BOX and DRAPER · Evolutionary Operation: A Statistical Method for Process Improvement
BOX, HUNTER, and HUNTER · Statistics for Experimenters: An Introduction to Design, Data Analysis, and Model Building
BOX and LUCEÑO · Statistical Control by Monitoring and Feedback Adjustment
BRANDIMARTE · Numerical Methods in Finance: A MATLAB-Based Introduction
BROWN and HOLLANDER · Statistics: A Biomedical Introduction
BRUNNER, DOMHOF, and LANGER · Nonparametric Analysis of Longitudinal Data in Factorial Experiments
BUCKLEW · Large Deviation Techniques in Decision, Simulation, and Estimation
CAIROLI and DALANG · Sequential Stochastic Optimization
CHAN · Time Series: Applications to Finance
CHATTERJEE and HADI · Sensitivity Analysis in Linear Regression
CHATTERJEE and PRICE · Regression Analysis by Example, *Third Edition*
CHERNICK · Bootstrap Methods: A Practitioner's Guide
CHERNICK and FRIIS · Introductory Biostatistics for the Health Sciences
CHILÈS and DELFINER · Geostatistics: Modeling Spatial Uncertainty
CHOW and LIU · Design and Analysis of Clinical Trials: Concepts and Methodologies
CLARKE and DISNEY · Probability and Random Processes: A First Course with Applications, *Second Edition*
*COCHRAN and COX · Experimental Designs, *Second Edition*
CONGDON · Bayesian Statistical Modelling
CONOVER · Practical Nonparametric Statistics, *Second Edition*
COOK · Regression Graphics
COOK and WEISBERG · Applied Regression Including Computing and Graphics
COOK and WEISBERG · An Introduction to Regression Graphics
CORNELL · Experiments with Mixtures, Designs, Models, and the Analysis of Mixture Data, *Third Edition*
COVER and THOMAS · Elements of Information Theory

*Now available in a lower priced paperback edition in the Wiley Classics Library.

COX · A Handbook of Introductory Statistical Methods
*COX · Planning of Experiments
CRESSIE · Statistics for Spatial Data, *Revised Edition*
CSÖRGŐ and HORVÁTH · Limit Theorems in Change Point Analysis
DANIEL · Applications of Statistics to Industrial Experimentation
DANIEL · Biostatistics: A Foundation for Analysis in the Health Sciences, *Sixth Edition*
*DANIEL · Fitting Equations to Data: Computer Analysis of Multifactor Data, *Second Edition*
DASU and JOHNSON · Exploratory Data Mining and Data Cleaning
DAVID · Order Statistics, *Second Edition*
*DEGROOT, FIENBERG, and KADANE · Statistics and the Law
DEL CASTILLO · Statistical Process Adjustment for Quality Control
DETTE and STUDDEN · The Theory of Canonical Moments with Applications in Statistics, Probability, and Analysis
DEY and MUKERJEE · Fractional Factorial Plans
DILLON and GOLDSTEIN · Multivariate Analysis: Methods and Applications
DODGE · Alternative Methods of Regression
*DODGE and ROMIG · Sampling Inspection Tables, *Second Edition*
*DOOB · Stochastic Processes
DOWDY and WEARDEN · Statistics for Research, *Second Edition*
DRAPER and SMITH · Applied Regression Analysis, *Third Edition*
DRYDEN and MARDIA · Statistical Shape Analysis
DUDEWICZ and MISHRA · Modern Mathematical Statistics
DUNN and CLARK · Applied Statistics: Analysis of Variance and Regression, *Second Edition*
DUNN and CLARK · Basic Statistics: A Primer for the Biomedical Sciences, *Third Edition*
DUPUIS and ELLIS · A Weak Convergence Approach to the Theory of Large Deviations
*ELANDT-JOHNSON and JOHNSON · Survival Models and Data Analysis
ETHIER and KURTZ · Markov Processes: Characterization and Convergence
EVANS, HASTINGS, and PEACOCK · Statistical Distributions, *Third Edition*
FELLER · An Introduction to Probability Theory and Its Applications, Volume I, *Third Edition*, Revised; Volume II, *Second Edition*
FISHER and VAN BELLE · Biostatistics: A Methodology for the Health Sciences
*FLEISS · The Design and Analysis of Clinical Experiments
FLEISS · Statistical Methods for Rates and Proportions, *Second Edition*
FLEMING and HARRINGTON · Counting Processes and Survival Analysis
FULLER · Introduction to Statistical Time Series, *Second Edition*
FULLER · Measurement Error Models
GALLANT · Nonlinear Statistical Models
GHOSH, MUKHOPADHYAY, and SEN · Sequential Estimation
GIFI · Nonlinear Multivariate Analysis
GLASSERMAN and YAO · Monotone Structure in Discrete-Event Systems
GNANADESIKAN · Methods for Statistical Data Analysis of Multivariate Observations, *Second Edition*
GOLDSTEIN and LEWIS · Assessment: Problems, Development, and Statistical Issues
GREENWOOD and NIKULIN · A Guide to Chi-Squared Testing
GROSS and HARRIS · Fundamentals of Queueing Theory, *Third Edition*
*HAHN and SHAPIRO · Statistical Models in Engineering

*Now available in a lower priced paperback edition in the Wiley Classics Library.

HAHN and MEEKER · Statistical Intervals: A Guide for Practitioners
HALD · A History of Probability and Statistics and their Applications Before 1750
HALD · A History of Mathematical Statistics from 1750 to 1930
HAMPEL · Robust Statistics: The Approach Based on Influence Functions
HANNAN and DEISTLER · The Statistical Theory of Linear Systems
HEIBERGER · Computation for the Analysis of Designed Experiments
HEDAYAT and SINHA · Design and Inference in Finite Population Sampling
HELLER · MACSYMA for Statisticians
HINKELMAN and KEMPTHORNE: · Design and Analysis of Experiments, Volume 1: Introduction to Experimental Design
HOAGLIN, MOSTELLER, and TUKEY · Exploratory Approach to Analysis of Variance
HOAGLIN, MOSTELLER, and TUKEY · Exploring Data Tables, Trends and Shapes
*HOAGLIN, MOSTELLER, and TUKEY · Understanding Robust and Exploratory Data Analysis
HOCHBERG and TAMHANE · Multiple Comparison Procedures
HOCKING · Methods and Applications of Linear Models: Regression and the Analysis of Variance, *Second Edition*
HOEL · Introduction to Mathematical Statistics, *Fifth Edition*
HOGG and KLUGMAN · Loss Distributions
HOLLANDER and WOLFE · Nonparametric Statistical Methods, *Second Edition*
HOSMER and LEMESHOW · Applied Logistic Regression, *Second Edition*
HOSMER and LEMESHOW · Applied Survival Analysis: Regression Modeling of Time to Event Data
HØYLAND and RAUSAND · System Reliability Theory: Models and Statistical Methods
HUBER · Robust Statistics
HUBERTY · Applied Discriminant Analysis
HUNT and KENNEDY · Financial Derivatives in Theory and Practice
HUSKOVA, BERAN, and DUPAC · Collected Works of Jaroslav Hajek—with Commentary
IMAN and CONOVER · A Modern Approach to Statistics
JACKSON · A User's Guide to Principle Components
JOHN · Statistical Methods in Engineering and Quality Assurance
JOHNSON · Multivariate Statistical Simulation
JOHNSON and BALAKRISHNAN · Advances in the Theory and Practice of Statistics: A Volume in Honor of Samuel Kotz
JUDGE, GRIFFITHS, HILL, LÜTKEPOHL, and LEE · The Theory and Practice of Econometrics, *Second Edition*
JOHNSON and KOTZ · Distributions in Statistics
JOHNSON and KOTZ (editors) · Leading Personalities in Statistical Sciences: From the Seventeenth Century to the Present
JOHNSON, KOTZ, and BALAKRISHNAN · Continuous Univariate Distributions, Volume 1, *Second Edition*
JOHNSON, KOTZ, and BALAKRISHNAN · Continuous Univariate Distributions, Volume 2, *Second Edition*
JOHNSON, KOTZ, and BALAKRISHNAN · Discrete Multivariate Distributions
JOHNSON, KOTZ, and KEMP · Univariate Discrete Distributions, *Second Edition*

*Now available in a lower priced paperback edition in the Wiley Classics Library.

JUREČKOVÁ and SEN · Robust Statistical Procedures: Aymptotics and Interrelations
JUREK and MASON · Operator-Limit Distributions in Probability Theory
KADANE · Bayesian Methods and Ethics in a Clinical Trial Design
KADANE AND SCHUM · A Probabilistic Analysis of the Sacco and Vanzetti Evidence
KALBFLEISCH and PRENTICE · The Statistical Analysis of Failure Time Data, *Second Edition*
KASS and VOS · Geometrical Foundations of Asymptotic Inference
KAUFMAN and ROUSSEEUW · Finding Groups in Data: An Introduction to Cluster Analysis
KEDEM and FOKIANOS · Regression Models for Time Series Analysis
KENDALL, BARDEN, CARNE, and LE · Shape and Shape Theory
KHURI · Advanced Calculus with Applications in Statistics, *Second Edition*
KHURI, MATHEW, and SINHA · Statistical Tests for Mixed Linear Models
KLUGMAN, PANJER, and WILLMOT · Loss Models: From Data to Decisions
KLUGMAN, PANJER, and WILLMOT · Solutions Manual to Accompany Loss Models: From Data to Decisions
KOTZ, BALAKRISHNAN, and JOHNSON · Continuous Multivariate Distributions, Volume 1, *Second Edition*
KOTZ and JOHNSON (editors) · Encyclopedia of Statistical Sciences: Volumes 1 to 9 with Index
KOTZ and JOHNSON (editors) · Encyclopedia of Statistical Sciences: Supplement Volume
KOTZ, READ, and BANKS (editors) · Encyclopedia of Statistical Sciences: Update Volume 1
KOTZ, READ, and BANKS (editors) · Encyclopedia of Statistical Sciences: Update Volume 2
KOVALENKO, KUZNETZOV, and PEGG · Mathematical Theory of Reliability of Time-Dependent Systems with Practical Applications
LACHIN · Biostatistical Methods: The Assessment of Relative Risks
LAD · Operational Subjective Statistical Methods: A Mathematical, Philosophical, and Historical Introduction
LAMPERTI · Probability: A Survey of the Mathematical Theory, *Second Edition*
LANGE, RYAN, BILLARD, BRILLINGER, CONQUEST, and GREENHOUSE · Case Studies in Biometry
LARSON · Introduction to Probability Theory and Statistical Inference, *Third Edition*
LAWLESS · Statistical Models and Methods for Lifetime Data, *Second Edition*
LAWSON · Statistical Methods in Spatial Epidemiology
LE · Applied Categorical Data Analysis
LE · Applied Survival Analysis
LEE and WANG · Statistical Methods for Survival Data Analysis, *Third Edition*
LePAGE and BILLARD · Exploring the Limits of Bootstrap
LEYLAND and GOLDSTEIN (editors) · Multilevel Modelling of Health Statistics
LIAO · Statistical Group Comparison
LINDVALL · Lectures on the Coupling Method
LINHART and ZUCCHINI · Model Selection
LITTLE and RUBIN · Statistical Analysis with Missing Data, *Second Edition*
LLOYD · The Statistical Analysis of Categorical Data
MAGNUS and NEUDECKER · Matrix Differential Calculus with Applications in Statistics and Econometrics, *Revised Edition*
MALLER and ZHOU · Survival Analysis with Long Term Survivors

MALLOWS · Design, Data, and Analysis by Some Friends of Cuthbert Daniel
MANN, SCHAFER, and SINGPURWALLA · Methods for Statistical Analysis of Reliability and Life Data
MANTON, WOODBURY, and TOLLEY · Statistical Applications Using Fuzzy Sets
MARDIA and JUPP · Directional Statistics
MASON, GUNST, and HESS · Statistical Design and Analysis of Experiments with Applications to Engineering and Science, *Second Edition*
McCULLOCH and SEARLE · Generalized, Linear, and Mixed Models
McFADDEN · Management of Data in Clinical Trials
McLACHLAN · Discriminant Analysis and Statistical Pattern Recognition
McLACHLAN and KRISHNAN · The EM Algorithm and Extensions
McLACHLAN and PEEL · Finite Mixture Models
McNEIL · Epidemiological Research Methods
MEEKER and ESCOBAR · Statistical Methods for Reliability Data
MEERSCHAERT and SCHEFFLER · Limit Distributions for Sums of Independent Random Vectors: Heavy Tails in Theory and Practice
*MILLER · Survival Analysis, *Second Edition*
MONTGOMERY, PECK, and VINING · Introduction to Linear Regression Analysis, *Third Edition*
MORGENTHALER and TUKEY · Configural Polysampling: A Route to Practical Robustness
MUIRHEAD · Aspects of Multivariate Statistical Theory
MURRAY · X-STAT 2.0 Statistical Experimentation, Design Data Analysis, and Nonlinear Optimization
MYERS and MONTGOMERY · Response Surface Methodology: Process and Product Optimization Using Designed Experiments, *Second Edition*
MYERS, MONTGOMERY, and VINING · Generalized Linear Models. With Applications in Engineering and the Sciences
NELSON · Accelerated Testing, Statistical Models, Test Plans, and Data Analyses
NELSON · Applied Life Data Analysis
NEWMAN · Biostatistical Methods in Epidemiology
OCHI · Applied Probability and Stochastic Processes in Engineering and Physical Sciences
OKABE, BOOTS, SUGIHARA, and CHIU · Spatial Tesselations: Concepts and Applications of Voronoi Diagrams, *Second Edition*
OLIVER and SMITH · Influence Diagrams, Belief Nets and Decision Analysis
PANKRATZ · Forecasting with Dynamic Regression Models
PANKRATZ · Forecasting with Univariate Box-Jenkins Models: Concepts and Cases
*PARZEN · Modern Probability Theory and Its Applications
PEÑA, TIAO, and TSAY · A Course in Time Series Analysis
PIANTADOSI · Clinical Trials: A Methodologic Perspective
PORT · Theoretical Probability for Applications
POURAHMADI · Foundations of Time Series Analysis and Prediction Theory
PRESS · Bayesian Statistics: Principles, Models, and Applications
PRESS · Subjective and Objective Bayesian Statistics, *Second Edition*
PRESS and TANUR · The Subjectivity of Scientists and the Bayesian Approach
PUKELSHEIM · Optimal Experimental Design
PURI, VILAPLANA, and WERTZ · New Perspectives in Theoretical and Applied Statistics

*Now available in a lower priced paperback edition in the Wiley Classics Library.

PUTERMAN · Markov Decision Processes: Discrete Stochastic Dynamic Programming
*RAO · Linear Statistical Inference and Its Applications, *Second Edition*
RENCHER · Linear Models in Statistics
RENCHER · Methods of Multivariate Analysis, *Second Edition*
RENCHER · Multivariate Statistical Inference with Applications
RIPLEY · Spatial Statistics
RIPLEY · Stochastic Simulation
ROBINSON · Practical Strategies for Experimenting
ROHATGI and SALEH · An Introduction to Probability and Statistics, *Second Edition*
ROLSKI, SCHMIDLI, SCHMIDT, and TEUGELS · Stochastic Processes for Insurance and Finance
ROSENBERGER and LACHIN · Randomization in Clinical Trials: Theory and Practice
ROSS · Introduction to Probability and Statistics for Engineers and Scientists
ROUSSEEUW and LEROY · Robust Regression and Outlier Detection
RUBIN · Multiple Imputation for Nonresponse in Surveys
RUBINSTEIN · Simulation and the Monte Carlo Method
RUBINSTEIN and MELAMED · Modern Simulation and Modeling
RYAN · Modern Regression Methods
RYAN · Statistical Methods for Quality Improvement, *Second Edition*
SALTELLI, CHAN, and SCOTT (editors) · Sensitivity Analysis
*SCHEFFE · The Analysis of Variance
SCHIMEK · Smoothing and Regression: Approaches, Computation, and Application
SCHOTT · Matrix Analysis for Statistics
SCHUSS · Theory and Applications of Stochastic Differential Equations
SCOTT · Multivariate Density Estimation: Theory, Practice, and Visualization
*SEARLE · Linear Models
SEARLE · Linear Models for Unbalanced Data
SEARLE · Matrix Algebra Useful for Statistics
SEARLE, CASELLA, and McCULLOCH · Variance Components
SEARLE and WILLETT · Matrix Algebra for Applied Economics
SEBER and LEE · Linear Regression Analysis, *Second Edition*
SEBER · Multivariate Observations
SEBER and WILD · Nonlinear Regression
SENNOTT · Stochastic Dynamic Programming and the Control of Queueing Systems
*SERFLING · Approximation Theorems of Mathematical Statistics
SHAFER and VOVK · Probability and Finance: It's Only a Game!
SMALL and McLEISH · Hilbert Space Methods in Probability and Statistical Inference
SRIVASTAVA · Methods of Multivariate Statistics
STAPLETON · Linear Statistical Models
STAUDTE and SHEATHER · Robust Estimation and Testing
STOYAN, KENDALL, and MECKE · Stochastic Geometry and Its Applications, *Second Edition*
STOYAN and STOYAN · Fractals, Random Shapes and Point Fields: Methods of Geometrical Statistics
STYAN · The Collected Papers of T. W. Anderson: 1943–1985

*Now available in a lower priced paperback edition in the Wiley Classics Library.

SUTTON, ABRAMS, JONES, SHELDON, and SONG · Methods for Meta-Analysis in Medical Research
TANAKA · Time Series Analysis: Nonstationary and Noninvertible Distribution Theory
THOMPSON · Empirical Model Building
THOMPSON · Sampling, *Second Edition*
THOMPSON · Simulation: A Modeler's Approach
THOMPSON and SEBER · Adaptive Sampling
THOMPSON, WILLIAMS, and FINDLAY · Models for Investors in Real World Markets
TIAO, BISGAARD, HILL, PEÑA, and STIGLER (editors) · Box on Quality and Discovery: with Design, Control, and Robustness
TIERNEY · LISP-STAT: An Object-Oriented Environment for Statistical Computing and Dynamic Graphics
TSAY · Analysis of Financial Time Series
UPTON and FINGLETON · Spatial Data Analysis by Example, Volume II: Categorical and Directional Data
VAN BELLE · Statistical Rules of Thumb
VIDAKOVIC · Statistical Modeling by Wavelets
WEISBERG · Applied Linear Regression, *Second Edition*
WELSH · Aspects of Statistical Inference
WESTFALL and YOUNG · Resampling-Based Multiple Testing: Examples and Methods for *p*-Value Adjustment
WHITTAKER · Graphical Models in Applied Multivariate Statistics
WINKER · Optimization Heuristics in Economics: Applications of Threshold Accepting
WONNACOTT and WONNACOTT · Econometrics, *Second Edition*
WOODING · Planning Pharmaceutical Clinical Trials: Basic Statistical Principles
WOOLSON and CLARKE · Statistical Methods for the Analysis of Biomedical Data, *Second Edition*
WU and HAMADA · Experiments: Planning, Analysis, and Parameter Design Optimization
YANG · The Construction Theory of Denumerable Markov Processes
*ZELLNER · An Introduction to Bayesian Inference in Econometrics
ZHOU, OBUCHOWSKI, and MCCLISH · Statistical Methods in Diagnostic Medicine

*Now available in a lower priced paperback edition in the Wiley Classics Library.